D0915868

On Evolution and Fossil Mammals

ON
EVOLUTION
AND
FOSSIL MAMMALS

BJÖRN KURTÉN

New York   COLUMBIA UNIVERSITY PRESS   1988

Library of Congress Cataloging-in Publication Data

Kurtén, Björn
On evolution and fossil mammals.

Bibliography: p.
Includes index.
1. Mammals, Fossil. 2. Evolution. I. Title.
QE881.K796 1988   569   86-32630
ISBN 0-231-05868-3

Columbia University Press
New York   Guildford, Surrey
Copyright © 1988 Columbia University Press
All rights reserved

Printed in the United States of America

CONTENTS

FOREWORD

by George Gaylord Simpson

Björn Kurtén is a paleontologist's paleontologist or, to modify a frequent comment of sports reporters, he may be characterized as a "world class" evolutionary biologist. He was born in Finland in 1924, and is now a professor of paleontology in a leading European university in the capital of Finland: Helsinki in Finnish, Helsingfors in Swedish.

Finland is a bilingual country generally listed among the Scandinavian countries. For historical reasons one of its languages is Swedish, one of the four distinct main Scandinavian languages, all of which are Indo-European and Germanic. The other official language of Finland, and the first language of most Finns, is, of course Finnish. It has virtually nothing in common with Swedish or with any other Indo-European language. There is some possible distant relationship of Finnish with Hungarian in a linguistic family called Finno-Ugrian. More distant and even more dubious is an early and remote linguistic spread from central Asia, where languages called Altaic are still spoken, into Europe, where Finnish and Hungarian persist as linguistic islands.

That digression on the subject of languages was written as background to the fact that Kurtén grew up to be multilingual. Almost all of his published writings are in Swedish, English, or both. Those reprinted in the present book were earlier published in English and are here reprinted in that language, which Kurtén speaks as well as writes fluently. At least seven of his books have been previously published in English, three of them published by Columbia University Press, one with an American co-author (Elaine Anderson). He has also published popular science, science fiction, and novels, mostly in Swedish.

In the author's note in one of his books in English, *The Cave Bear Story* (Columbia University Press, 1976) Kurtén told how he became enticed by, one might say entangled in, paleontology—especially mammalian and especially in the Pleistocene, the last geological epoch before the Holocene or Recent. He wrote:

"No one escapes his fate. It might be said that my affair with the cave bear started half a century ago when it was decided to give the child a name [Björn] that happens to be Swedish for bear. There were some early difficulties in living up to it. . . . Still the real thing began in the early 1950s. Eager to apply newfangled population ideas on fossil mammals, I was casting about for a statistically respectable sample of some fossil mammal—any fossil mammal. The one that happened to be at hand had been collected a hundred years earlier, and had been lying, more or less forgotten, for many decades in the cupboards of the geology department of the University of Helsinki. It was the cave bear: hundreds and hundreds of teeth and bones."

The man named Bear (Björn), no longer a child, later extended his observations on *Ursus spelaeus,* the cave bear, throughout Europe. He also extended his study, both geographical and evolutionary, to the whole family of bears, the Ursidae. This was one of the last distinct families of mammals to arise by evolution. It eventually spread to all the continents except Australia and Antarctica and all climates. Kurtén eventually studied practically all known members of that family, both fossil and living.

Sweden, close to Helsingfors and where Kurtén's first language was spoken, became almost a second home, or at least a working base for him. (It happens that although I have never been in Finland I have met Björn elsewhere.) In the midst of his other travels and interests, Kurtén established a further basis for "newfangled population ideas on fossil mammals" in Sweden. A main center for paleontology in Sweden has long been in the Naturhistoriska Riksmuseet in the outskirts of Stockholm. The major paleontological research interest there has been on primitive fishes, following the lead of Erik Stensiö, who lived and continued to work there until his death at a great age, in January 1984. In Uppsala, a smaller city some fifty miles north-northwest of Stockholm, Kurtén found more materials relevant to his main interests. There is the university where Linnaeus, who provided the keystone of taxonomic natural history, had been a professor. There is a large (long post-Linnaean) collection of fossil mammals. Most of these are from Mongolia and China and in age range from mid to late Cenozoic (the latest of the major eras of geology and of evolution). Most of that collection was made by a Sino-Swedish expedition, the last project organized by Sven Hedin (1865–1952), a famous Swedish explorer in Asia.

Parts of the collections made under those auspices were studied in Uppsala by Birger Bohlin from the 1930s into the early 1950s. As Bohlin ceased to work over that collection, Kurtén took parts of it over. Some of his results are evident in the long paper "On the Variation and Population Dynamics of Fossil and Recent Mammal Populations," reprinted in the present book (article 1). Then and thereafter, Kurtén also studied many kinds of fossils; but his main interest was always mammals, both fossil and recent, in museums and other collections from western Russia through Great Britain and from northern Scandinavia to the Mediterranean and Africa. At various times he likewise visited and studied almost all important collections in North America. Such extension and versatility are further exemplified, at least briefly, by Kurtén's highly original orientation in quantification, evolution, biogeography, and other approaches to zoological and more general biological subjects in the papers reprinted here.

This sampling of his papers, selected and annotated by Kurtén, covers many of his ideas and methods as they first occurred to him. The reprinting is largely to make available some of the original journal papers not otherwise now generally in print. It will also be helpful here to mention some of his books, mostly more popular in style than his technical papers and always embodying his wide acquaintance both with a variety of subjects and with other studies in the fields covered by Kurtén, cited in the following mini-reviews of books by him:

1968. *Pleistocene Mammals of Europe*. Aldine Publishing Co. A thorough, mostly first-hand account that Kurtén has studied at great length starting from the beginning of his career.

1972. *Not from the Apes*. Pantheon Books (Random House). Not Kurtén's major field and in some respects unorthodox now, but a useful and original discussion of the origin and evolution of our own species.

1972. *The Age of Mammals*. Columbia University Press. A fairly simple but extensive review of the animals, mammals and a few others, during this long—about 65 million years—era of the earth's history. It was meant as a companion volume to Colbert's *The Age of Reptiles* (1965) and Osborn's now obsolescent book on nearly the same subject in

1910. Kurtén's book has many life restorations, attractive but a few with perhaps a bit too much artistic license.

1976. *The Cave Bear Story*. Columbia University Press. Both highly competent and pleasantly readable. A quotation from the "Author's Note" was given at the beginning of this introduction.

1980. *Pleistocene Mammals of North America*. (With Elaine Anderson.) Columbia University Press. A compendious and detailed, somewhat technical compilation with locality maps and some other figures. Like any book of this sort it will remain useful for years, but will not always be the last word.

I also cannot forbear including one other book, which is currently available only in Swedish, but of which there was a now practically unobtainable English edition:

1972. *The Ice Age*. Putnam's Sons. This small English edition quickly sold out and has not been reprinted. It was translated from the 1969 Swedish edition, *Istiden,* printed in Italy, published by "Forum." This book is especially noteworthy for its fine illustrations, mostly in color.

G. G. Simpson

PREFACE

The majority of the papers reprinted here appeared originally in the journals *Acta Zoologica Fennica* and *Commentationes Biologicae,* which were then published by the Societas pro Fauna et Flora Fennica and Societas Scientiarum Fennica, respectively, between 1953 and 1973. (The *Commentationes* have since been discontinued, while the *Acta* is now published jointly by both societies and two additional learned societies, the Finnish Academy of Sciences and the Vanamo Biological Society, all in Helsinki.) Although both journals had a limited distribution and the issues are now mostly out of print and hence difficult to obtain, the papers have continued to be in demand by students of vertebrate paleontology and related subjects. That, of course, is the *raison d'être* for this book, which contains, in addition to these hard-to-obtain articles, some papers published elsewhere, included to fill out the picture. I wish to record my gratitude to the copyright holders as credited.

The aim is to present certain concepts and methodological innovations as first conceived and published, and in some instances to follow their later development. It goes without saying that some of the conclusions formed at the time have turned out to be premature and have had to be revised in the light of later research (for instance, as regards absolute chronology and taxonomy).

Paleotheriology (surely a better term than the hybrid "paleomammalogy") has a long tradition of using the genus as the basic category. Today, that must be seen as a hangover from a bygone age; I have always been convinced that the species should play that role, as it does in neozoology. (In a study of extinction, for instance, the disappearance of a monotypic genus would loom larger than that of 9 out of 10 species in another genus.) Thus, good species- and population-level work is the ultimate basis for a good taxonomy and all studies relying thereon. This is perhaps the single and earnest intent (alas, not always realized) running through this heterogeneous collection.

What is a paleospecies? Others have offered definitions; I desist. We have, of course, to work with morphospecies, and the approach has to be pragmatic. We use "qualitative" and "quantitative" characters. At present, although many studies of broader phylogenetic and other problems are being carried out with sophisticated mathematical techniques (with excellent results), much of the work around the species level remains essentially qualitative, substituting crude guesswork for appropriate metric study. And yet the appropriate tools were presented to vertebrate paleontologists and zoologists almost half a century ago (Simpson and Roe 1939).

What is wrong: Computer envy? It may be felt that, with all this tremendous hardware at hand, why do simple things yourself, when you can get an expert to do complicated ones? The point is that quite elementary statistical procedures, applied with imagination and understanding, will go a long way toward solving many problems. As an example, multivariate statistics have many important uses. Still, Medawar (1977) properly warns against the "illusion embodied in the ambition to attach a single number valuation to complex quantities." A multivariate study of a whole dentition is very useful, and may enable us to discriminate between two species; but it will not tell us that species A has a longer talonid in M_1 than species B, and still less lead us to ask why. I might mention, too, the problem of the loss of information resulting from the necessary scrapping of all kinds of incomplete material—such as isolated teeth—which will always be the common stock-in-trade of paleotheriology.

We have many tools. Some of them are trendy just now; but we need them all. Flying altitude should be adapted to the problem at hand.

A few words about the subjects of these papers. The first and longest paper treats, among other things, with population dynamics as reflected in fossil assemblages. My own introduction to the topic came in 1951 at the University of Helsinki, when I attended a lecture series by Dr. Olavi Kalela, an enthusiastic and brilliant mammalogist. At the same time, I was playing around with certain data on worn bovid molar crown heights published years earlier by Bohlin (1935), and to my amazement found that they tended to cluster around certain values, apparently representing annual age groups. Obviously, this was something to go after: but I was in Finland,

and the material was in Uppsala, Sweden; and at that time international travel was severely restricted (my country was then paying War reparations and everything else was secondary, including budding vertebrate paleontologists with crazy ideas in their heads). In fact I had to fret for about a year, waiting for a chance to get to Uppsala and see if this would work, as indeed it turned out to do. Professor Pontus Palmgren, my teacher in zoology and the most inspiring teacher anybody could wish for, told me, "You have to rush into print, or else somebody else will get the same idea; it's in the air." I did rush into print (Kurtén 1953) but it took a long time to catch on, so apparently it was not quite that much in the air. He was prodding me, of course. May every student have such a mentor.

The topic has been intermittently treated by a few authors since the publication later in 1953 of the main paper, reprinted here (e.g. Kurtén 1954a, 1954b, 1958, 1964, 1976, 1983; Van Valen 1964; Voorhies 1969, a seminal paper) and has recently become the focus of a new surge of interest (Hulbert 1982; Klein, 1982; Klein and Cruz-Uribe 1983).

The study of natural selection by differential mortality affecting dental dimensions, which was introduced in the same paper, has been the subject of various later publications (Kurtén 1954a, 1954b, 1955, 1957, 1958, 1964, 1967, 1976, 1983; Van Valen 1963, 1965a, 1965b; Van Valen and Weiss 1966; Perzigian 1975; Wolpoff 1976). Analogous methods can be used for invertebrates (Sambol and Finks 1977, on bivalves).

A third topic introduced in the same paper, the correlation fields in dentition, has received little attention, although the work has been lightened by the advent of the computer; later studies include Van Valen (1962), Suarez and Pernor (1972), Lavelle (1978), Zingeser and Phoenix (1978), Gingerich and Winkler (1979).

A number of these papers deal with allometry. I was much interested to find that intrapopulation allometry in dentitions could be used in various ways in the study of evolution on the population level (article 3) and that "matching" of allometry might clarify relationships between populations (article 4). In this section is included an interesting case of evolutionary reversal (article 5). Article 6 is intended to show some ways in which the study of rates of migration may shed light on paleobiogeographical problems, and also to demonstrate the fallacy of a provincial taxonomy.

I have always accepted continental drift as a fact, and when Richard Carrington asked me to write a sequel to Colbert's (1965) *The Age of*

Reptiles, I felt it necessary to look into the problem (see Kurtén 1971). Articles 7 and 8 reflect this interest. The subject now belongs to the mainstream of paleontology and the literature is rapidly growing. As outstanding contributions on fossil mammals I cite McKenna (1973, 1975, 1983).

Many methods of assessing turnover rates in faunas have been proposed. The half-life method treated here (articles 9 and 10) has the merit of simplicity.

Paleoethological conclusions based on anatomical traits may be of great interest (see, e.g., Thenius 1971). Guthrie (1970, 1977) studied "body hot spots" in modern man; in the brief final paper (article 11), the theme is pursued with regard to fossil man.

In retrospect, the shortcomings of many of these papers are all too evident to me. However, as Birger Bohlin once told me, "if you don't make any mistakes, you won't make any 'takes'. " And so we should cheerfully go on, sticking our necks out.

I wish to thank George Gaylord Simpson for his foreword. My own work, like that of evolutionary paleontology as a whole, stems directly from his *Tempo and Mode in Evolution* (1944), which marks the great turning point in paleontological thinking.

Björn Kurtén

REFERENCES

Bohlin, B. 1935. Cavicornier der Hipparion-Fauna Nord-Chinas. *Palaeontologia Sinica,* Ser. C, vol. 9, Fasc. 4: 1–166.

Colbert, Edwin H. 1965. *The Age of Reptiles.* London: Weidenfeld and Nicolson, 228 p.

Gingerich, P. and D. A. Winkler. 1979. Patterns of variation and correlation in the dentition of the red fox *(Vulpes vulpes). J. Mammal.* 60: 691–704.

Guthrie, R. D. 1970. Evolution of human threat display organs. *Evolutionary Biology* 4: 257–302.

——. 1977. *Body Hot Spots.* New York: Kangaroo Books.

Hulbert, R. C., Jr. 1982. Population dynamics of the three-toed horse *Neohipparion* from the late Miocene of Florida. *Paleobiology* 8: 159–167.

Klein, R. G. 1982. Age (mortality) profiles as a means of distinguishing hunted species from scavenged ones in Stone Age archeological sites. *Paleobiology* 8: 151–158.

Klein, R. G. and K. Cruz-Uribe. 1983. The computation of ungulate age (mortality) profiles from dental crown heights. Paleobiology 9: 70–78.

Kurtén, B. 1953. Age groups in fossil mammals. *Commentationes Biologicae* 13 (13): 1–6.

——. 1954 a. Population dynamics and evolution. *Evolution* 8: 75–81.

——. 1954 b. Population dynamics—a new method in paleontology. *J. Paleontol.* 28: 286–292.

——. 1955. Sex dimorphism and size trends in the cave bear, *Ursus spelaeus* Rosenmüller and Heinroth. *Acta Zool. Fennica* 90: 1–48.

——. 1957. A case of Darwinian selection in bears. *Evolution* 11: 412–416.

——. 1958. Life and death of the Pleistocene cave bear, a study in paleoecology. *Acta Zool. Fennica* 95: 1–59.

——. 1964. Population structure in paleoecology. In J. Imbrie and N. D. Newell, eds. *Approaches to Paleoecology;* pp. 91–106. New York: John Wiley and Sons.

——. 1967. Continental drift and the palaeogeography of reptiles and mammals. *Commentationes* B.

——. 1971. *The Age of Mammals*. London: Weidenfeld and Nicholson; New York: Columbia University Press.

——. 1976. *The Cave Bear Story*. New York: Columbia University Press.

——. 1983. Variation and dynamics of a fossil antelope population. *Paleobiology* 9: 62–69.

Lavelle, C. L. B. 1978. Correlations between tooth dimensions of man and apes. *Acta Anatomica* 102: 358–364.

McKenna, M. C. 1973. Sweepstakes, filters, corridors, Noah's arks, and beached Viking funeral ships in palaeogeography. In D. H. Tarling and S. K. Runcorn, eds. *Implications of Continental Drift to the Earth Sciences,* 1: 295–308. New York: Academic Press.

——. 1975. Fossil mammals and early Eocene North Atlantic land continuity. *Ann. Missouri Bot. Garden* 62: 335–353.

——. 1983. Cenozoic paleogeography of North Atlantic land bridges. In Bott, et al. eds. *Structure and Development of the Greenland-Scotland Ridge,* pp. 351–399. New York: Plenum.

Medawar, P. B. 1977. Unnatural science. New York Review of Books, February 3; pp. 13–18.

Perzigian, A. J. 1975. Natural selection in the dentition of an Arikara population. *Amer. J. Phys. Anthropol.* 42: 63–70.

Sambol, M. and R. M. Finks. 1977. Natural selection in a Cretaceous oyster. *Paleobiology* 3: 1–16.

Simpson, G. G. 1944. *Tempo and Mode in Evolution*. New York: Columbia University Press.

Simpson, G. G. and A. Roe. 1939. *Quantitative Zoology*. New York: McGraw-Hill. 414 p.

Suarez, B. K. and R. Pernor. 1972. Growth fields in the dentition of the gorilla. *Folia Primatologica* 18: 356–367.

Thenius, E. 1971. Sozialverhalten vorzeitlicher Schweine. *Umschau* 7/1971: 248.

Van Valen, L. 1962. Growth fields in the dentition of *Peromyscus*. *Evolution* 16: 272–277.

——. 1963. Selection in natural populations: *Merychippus primus,* a fossil horse. *Nature* 197: 1181–1183.

——. 1964. Age in two fossil horse populations. *Acta Zoologica* 45: 93 – 106.

——. 1965 a. Selection in natural populations. III. Measurement and estimation. *Evolution* 19: 514 – 528.

——. 1965 b. Selection in natural populations. IV. British house mice *(Mus musculus)*. *Genetica* 36: 119 – 134.

Van Valen, L. and R. Weiss. 1966. Selection in natural populations. V. Indian rats *(Rattus rattus)*. *Genet. Bes. Cambridge* 8: 261 – 267.

Voorhies, M. R. 1969. Taphonomy and population dynamics of an early Pliocene vertebrate fauna, Knox County, Nebraska. *Contr. Geol. Univ. Wyoming, Spec. Papers* 1: 1 – 69.

Wolpoff, M. H. 1976. Australopithecine tooth size selection. *Amer. J. Phys. Anthropol.* 44: 216.

Zingeser, M. R. and C. H. Phoenix. 1978. Materic characteristics of the canine dental complex in prenatally androgenized female rhesus monkeys *(Macaca mulatta)*. *Amer. J. Phys. Anthropol.* 49: 187 – 192.

I
POPULATION DYNAMICS AND SELECTION

ONE

On the Variation and Population Dynamics of Fossil and Recent Mammal Populations

INTRODUCTION: "SERO VENIENTIBUS OSSA"

The paleontologist, indeed, generally comes much too late to find anything but bones. Instead he finds something denied to the neontologist: the time element; and the crowning biological achievement of paleontology has been the demonstration, from the history of life, of the validity of the evolution theory. If . . . we turn to the record of geologic history to answer the question whether as a matter of fact the diverse types of existing animals did evolve from common primitive stocks, or whether the various races have remained unchanged since they were created, the answer is perfectly definite and categorical. Whenever the evidence is sufficient . . . it proves the theory to be a fact of record. (Matthew 1926).

"THE EVIDENCE includes *a fortiori* a demonstration of the course of evolution—whenever the evidence is sufficient." This historical record, therefore, provides a check on the various theories of evolutionary mechanics: a theory not consistent with the historical data cannot be passed. Here, however, according to some authors, the contribution of paleontology must perforce end. "Paleontology is of such a nature that its data by themselves cannot throw any important light on genetics or selection."

Reprinted, with permission, from *Acta Zoologica Fennica* (1953) 76:1–122.

(Huxley 1942). No doubt this (in my opinion) somewhat rash generalization—provoked, it would seem, by understandable irritation at the Lamarckist and orthogenetic views held by many earlier paleontologists—may be expected to induce a defeatist attitude among evolution-minded paleontologists. Fortunately this attitude is not all-pervading, as amply shown, e.g., by Simpson's epoch-making work (1944). My personal opinion is that paleontology, and especially paleoecology, may yet make new important contributions to the study of evolution, and that paleontological data may indeed throw important light in particular on selection. The present work is intended rather to demonstrate the possibility than to fulfill any part of the prophecy; to point out, as it were, that the challenge of the fact *sero venientibus ossa* is most fittingly met by a quest for new tools of study.

The approach to the subject and the points of view adopted here are mainly new to paleontology, though based on well-known ecological and genetic branches of study. The present application has necessitated the development of some new methods of study. I am well aware that the techniques employed may be found lacking in many respects. I also wish to stress that in most cases the data at hand are still far from numerous enough and that the interpretations may be biased in various ways. As an apology I can only quote Simpson, who, though certainly with less cause for hesitation, observes: "It is, however, pusillanimous to avoid making our best efforts today because they may appear inadquate tomorrow . . . The data will never be complete, and their useful, systematic acquisition is dependent upon the interpretation of the incomplete data already in hand."

The main part of the data to be discussed in this thesis were obtained from studies on the Chinese *Hipparion* fauna of the Lagrelius collection, Uppsala, and I am deeply indebted to Professor P. Thorslund for permission to study the material and to use the technical apparatus of the Paleontological Institute.

The Lagrelius collection has been most competently described by Bohlin, Hopwood, Pearson, Ringström, Sefve, and Zdansky, under the leadership of the late Professor C. Wiman. In 1952 I published a quantitative study of the Pontian part of the collection and discussed the general character of the environment. The treatment revealed, as had also been indicated by Schlosser (1901) and Bohlin (1935), that there was a northwestern fauna (*dorcadoides* fauna) of plains type, and a southeastern one (*gaudryi* fauna) of forest type. It is also suggested that there was an intermediate biotope, populated in part by members of the two main faunas, and in part, perhaps, by forms peculiar to this zone (notably, *Gazella paotehensis*).

Many species are represented by a very rich material, and the present paper will be concerned with some of these. The most important large samples are of the genera *Ictitherium, Crocuta, Hipparion, Dicerorhinus, Chilotherium, "Diceratherium," Chleuastochoerus, Cervocerus, Samotherium, Urmiatherium, Plesiaddax,* and *Gazella.* Many of these are represented by several species. The sample most frequently discussed is that of the Hyaenidae, and a reconsideration of the status of the hyaenid populations has therefore been added.

The methodological basis for the treatment is in principle very simple: it is the mensuration of mammalian tooth crowns in two dimensions—length (or occasionally breadth) and height. The former is the standard measurement of size; the latter yields a relative estimate of individual age, which, under certain favorable circumstances, may be transformed into a measure of absolute age. Various possibilities of analytic and synthetic treatment of these data will be explored. The first part of the thesis concerns the variation and correlation of teeth in mammals, the second part the age structure of the populations. Introduction to these subjects is deferred to the opening chapters of Parts I and II.

Mensuration was generally carried out with a caliper. Special care was taken to ensure consistency. For each individual population the same implement was used throughout, and the mensuration was checked after some lapse of time. Thus the possible bias, as regarding both implement and method, should be consistent, and its damaging effects eliminated. The statistical text-books mainly employed were Simpson and Roe (1934) and Pearl (1940). Most calculations were carried out with a large slide rule (50 cirs. giving correct values to 3–4 places. This ensures accuracy in calculation, which is desirable even where the results are subsequently rounded.

———

I wish to address my thanks to my teacher in zoology, Professor Pontus Palmgren, Helsingfors, for his constructive criticism, encouragement and advice. Also I am greatly indebted to Dr. Olavi Kalela, Custodian of Helsingfors University Zoological Museum, for his kind interest and valuable advice, and for permission to study the material of the Museum. My thanks are also due to Professor Sven Hörstadius, Uppsala University Zoological Institute, for permission to study the skull material of the Museum.

For valuable data and advice I am further indebted to Drs. Otto Zdansky, Birger Bohlin, Valdar Jaanusson, and Björn Petersen, Uppsala, and Drs. Lars von Haartman, Walter Hackman, and Mr. Henrik Wallgren, M. A., Helsingfors. I wish also to extend my thanks to students abroad, in particular to Dr. R. A. Stirton, University of California, and Dr. G. G. Simpson, Columbia University, who have facilitated my work by providing me with literature the access to which would otherwise have been difficult.

Mr. R. Washbourn (The British Council, Helsingfors) has very kindly read the English text.

This study has been aided by grants from Svenska vetenskapliga centralrådet, Helsingfors, and from Vasa Nation at Helsingfors University.

ON THE CORRELATION IN THE MAMMALIAN DENTITION

Characteristics of the Correlation

In their studies of correlation in the human hand, Pearson and his collaborators (see Whiteley and Pearson 1900; Lewentz and Whiteley 1902) stated the "rule of neighborhood"—i.e., that adjacent parts of an organ are more strongly intercorrelated as to size than more distant parts. This rule has since then been confirmed for many other cases, e.g., the appendages of the crayfish (Pearl and Clawson 1907); phalanges of the foot in crows, fowl, and ducks (Alpatov and Boschko-Stepanenko 1928); joints of antennae in insects (ibid). The latter authors made also another generalization, viz., that correlation is higher between proximal parts of an organ than in distal ones.

For the most part, studies of this kind have been concerned with organs the absolute size of which is changed during growth. Now this will inevitably introduce a more or less spurious correlation, as in the case of the crayfish (Pearl and Clawson), where the sample allegedly comprised individuals in different moults; the coefficients of correlation are, accordingly, very high (in general $r > .9$). The study by Alpatov and Boschko-Stepanenko on the antennal joints of *Pyrrhocoris apterus* avoids spurious correlation by being confined to adult specimens only and taking the sexes separately.

Study of this kind might profitably be directed toward the dimensions of mammalian tooth crowns. The problem of spurious correlation due to growth is not posed by mammalian dentitions (with very rare exceptions), and such studies would seem to hold especial interest to the paleontologist. This possibility has however been largely neglected, except for some isolated calculations of correlation coefficients.

Thus Simpson and Roe (1939) give the coefficient of correlation for the size of two adjacent teeth (the first and second lower molars of *Phenacodus primaevus*). In this case, $r = .82$, a value which is fairly common for variates of this kind.

I have carried out a more extensive study of the correlations in the mammalian dentition in order to investigate somewhat more closely the laws of variation in these organs. By such means it seems possible to attain some insight into the connection between function, structure, and genetics of the teeth. The following treatment comprises some examples of this correlation, a discussion of the causes underlying it, and an instance of its evolutionary change.

Since the number of observation pairs was relatively low in each case, grouping has been avoided, and correlation was calculated by taking actual sums of the squares and of the products of deviations. The labor involved is considerable and might well in future be reduced to some extent by use of the method described by Miller and Weller, 1952. It was however felt that in this initial study as accurate results as possible should be aimed at. For great help in this work I am indebted to my wife, Ruth M. Kurtén.

CORRELATION IN CHLEUASTOCHOERUS MOLARS, AND ITS IMITATION BY MEANS OF MODELS

For the Pontian Chinese Pig, *Chleuastochoerus stehlini,* the correlation coefficients for the upper molars were calculated by me from Pearson's (1928) data, considering the length of the tooth crowns. The result was:

Correlation between M^1 and M^2: $r = .83$ (N = 25)
Correlation between M^2 and M^3: $r = .78$ (N = 21)
Correlation between M^1 and M^3: $r = .62$ (N = 18)

Thus the rule of neighborhood is again confirmed. The correlation is positive throughout, but that between adjacent teeth is higher than that between the first and third molars.

At first sight there does not seem to be any *a priori* functional reason for the fact that adjacent teeth are positively correlated as to size, except perhaps for some advantage in the maintenance of a certain size ratio to each other of the different functional parts of the dentition. But it might equally well be argued, from a functional point of view, that a negative correlation between adjacent teeth could be of advantage, so that a tooth smaller than mean might be compensated functionally by a larger tooth next adjacent. This would be in accord with the "Kompensationsprinzip" (Rensch 1947), but this principle does not hold in the present case.

The causes of this correlation are probably genetic, and its peculiar property of decreasing with increasing distance points clearly to some kind of field action. The conclusion would be that the tooth row is controlled by a genetic "growth field," constituted by a set of polymeric genes. Each of these genes or gene groups would exert a major influence in part of the field, but also affect nearby parts, and the ranges of the separate genes would permeate each other and merge together. Thus, one gene might act primarily on M^3, causing this tooth to deviate in a certain direction from the mean, but part of its influence would extend to M^2 and M^1, with successively decreasing effect.

The principle may be illustrated by means of a model (table 1.1). It is assumed that the size of three successive teeth (fig. 1.1a, I—III) is determined by two gene pairs with alleles Aa and Bb. The minimum size of these teeth is assumed to be 10 mm for each. Alleles a and b do not add to the size of the teeth. A, on the other hand, if present, effects an increase of 2 mm in the length of (I), 1.5 mm in the length of (II), and 1 mm in the length of (III). B, in turn, brings about an increase of 2 mm in the length of (III), 1.5 mm in the length of (II) and 1 mm in the length of (I). The possible combinations, resulting crown lengths, and frequencies appear from the table. The resulting correlations are:

$$(I)— (II): r = .95$$
$$(II)—(III): r = .95$$
$$(I)—(III): r = .80$$

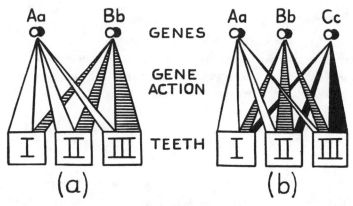

Fig. 1.1 Diagrammatic representation of gene action on tooth crown size, to explain correlation between adjacent teeth and the "rule of neighbourhood." (a), with two pairs of alleles; resulting correlation high; (b), with three pairs of alleles; resulting correlation moderate.

Table 1.1
Combinations and Effects on Tooth Crown Size of Two Pairs of Alleles (Model)
(as in Fig. 1.1a)

Gene		Increase (A)			Increase (B)			Resulting size		
Combinations	f	(I)	(II)	(III)	(I)	(II)	(III)	(I)	(II)	(III)
aabb	1	—	—	—	—	—	—	10	10	10
AAbb	1	4	3	2	—	—	—	14	13	12
Aabb	2	2	1.5	1	—	—	—	12	11.5	11
aaBb	2	—	—	—	1	1.5	2	11	11.5	12
AaBb	4	2	1.5	1	1	1.5	2	13	13	13
AABb	2	4	3	2	1	1.5	2	15	14.5	14
AaBB	2	2	1.5	1	2	3	4	14	14.5	15
aaBB	1	—	—	—	2	3	4	12	13	14
AABB	1	4	3	2	2	3	4	16	16	16

f, frequency. Increase (A) and (B) indicates the length increase (in mm.) from minimum (10 mm.) added by the action of allele A or B. Resulting size in mm.

The pattern is thus similar to that in the actual case of *Chleuastochoerus;* but the correlation is much higher, possibly because the number of allele pairs in reality is higher than two per three teeth.

A very close approximation to conditions in *Chleuastochoerus* is attained if the pairs of alleles are increased to three (fig. 1.1b), and their action on the length of the tooth crowns is assumed to be as follows:

Tooth	Increment in crown length, caused by gene		
	A	B	C
(I)	4	2	1
(II)	2	4	2
(III)	1	2	4

The resulting combinations and crown lengths are given in table 1.2. The correlations are as follows:

$$(I)—(\ II: r = .83$$
$$(II)—(III): r = .83$$
$$(I)—(III): r = .66$$

The correlations between the *Chleuastochoerus* molars do not differ significantly from these; consequently it may be stated that one pair of alleles per tooth is enough to account for the data. *Essentia non sunt multiplicanda praeter necessitatem.*

The genotypic series of the length of (I) and (III) is divided into 14 steps and that of (II) into 9 steps. When phenotypic variation is added, as well as the action of genes governing other parts of the tooth row, the resulting variation of crown size will be completely continuous. The model should not however be overinterpreted; the actual causal chain may be vastly more complicated. Indeed, this may be a problem for phenogenetics rather than formal genetics.

CORRELATION BETWEEN THE MOLARS OF HIPPOPOTAMUS AMPHIBIUS

The correlation between the molars in both jaws of *Hippopotamus emphibius* L. was studied on the basis of measurements recorded by Hooijer

Table 1.2
Combinations and Effects on Tooth Crown Size of Three Pairs of Alleles
(Model)
(as in Fig. 1.1b)

Gene Combinations	Frequency	Resulting size of		
		M^1	M^2	M^3
aabbcc	I	10	10	10
AAbbcc	I	18	14	12
aaBBcc	I	14	18	14
aabbCC	I	12	14	18
Aabbcc	2	14	12	II
aaBbcc	2	12	14	12
aabbCc	2	II	12	14
AABbcc	2	20	18	14
AAbbCc	2	19	16	16
AaBBcc	2	18	20	15
AaBbcc	4	16	16	13
AabbCc	4	15	14	15
aaBbCc	4	13	16	16
AaBbCc	8	17	18	17
(Mean)				
AABbCc	4	21	20	18
AaBBCc	4	19	22	19
AaBbCC	4	18	20	21
aaBBCc	2	15	20	18
aaBbCC	2	14	18	20
AabbCC	2	16	16	19
AABBCc	2	23	24	20
AABbCC	2	22	22	22
AaBBCC	2	20	24	23
AABBcc	I	22	22	16
AAbbCC	I	20	18	20
aaBBCC	I	16	22	22
AABBCC	I	24	26	24

(1950a). Both sexes were represented in the sample. The females average slightly smaller than the males; the difference, though of doubtful significance, may introduce a spurious, intergroup correlation, if the coefficients are based on the sample as a whole. On the other hand, the subsamples (22 males and 10 females) are somewhat too small to give reliable results when considered separately. The difficulty was resolved by computing separately the means for the females and for the males, and the correlation coefficients were found from the formula,

$$(i) \qquad r = \frac{\Sigma\,(d_{x\delta}\,d_{y\delta}) + \Sigma\,(d_{x\varphi}\,d_{y\varphi})}{\sqrt{[\Sigma\,(d_{x\delta}^2) + \Sigma\,(d_{x\varphi}^2)][\Sigma\,(d_{y\delta}^2) + \Sigma\,(d_{y\varphi}^2)]}}$$

It is thus assumed that conditions are identical or closely similar in both sexes; or, in terms of the model, that homologous genes act on tooth size (thus possible sex-linked genes are disregarded). The results are summarized in table 1.3. The correlations within the upper and lower dentitions, respectively, show a pattern basically similar to that in *Chleuastochoerus,* but it can be seen that the coefficients of correlation are markedly lower in the present instance. As the *Chleuastochoerus* sample is not homogeneous as to sex, and may indeed be heterogeneous taxonomically (the facts suggest, but do not prove, subspecific differentiation; see Helga Pearson, 1928), this is not unexpected. The model of table 1.2 will apply to the present data, if the action of the genes is assumed to be somewhat more restricted than in the demonstration. The rule of neighborhood holds good, the correlation between first and third molars always being lower than that between adjacent molars; indeed, r_{M1M3} is of doubtful significance.

In the correlation between teeth in upper *and* lower jaws some remarkable features are noted. In the first place it can be seen that the rule of neighborhood is still valid, the strongest correlation invariably existing between occluding teeth, and the values falling with increasing distance. But it appears that the correlation between occluding teeth is much higher throughout than that between adjacent teeth in one of the jaws. If these

Table 1.3
**Coronal Length correlation between Molars of *Hippopotamus amphibius* L.,
25 Skulls, Males and Females Combined; N from 10 to 25. Raw Data from
Hooijer. Table of *r*.**

M^2	M^3				M_2	M_3	
.56	.53	M^1			.59	(.30)	M_1
	.71	M^2				.52	M_2

M^1	M^2	M^3	
.76	.53	(.55)	M_1
.55	.87	.65	M_2
(.45)	.55	.72	M_3

( ) not significant

properties are governed in the way suggested in the model, this would imply that, for instance, $M\ 2/2$ have more genes in common ($r = .87$) than do M^1—^{2} ($r = .56$). The impression is indeed that the correlation fields of the upper and lower tooth rows are identical or nearly so: for the values of r are almost always closely similar for a given part of the dentition, whether they represent correlation within one jaw or between upper and lower teeth. Thus:

$$^rM^1M^2 = .56$$
$$^rM_1M_2 = .59$$
$$^rM^1M_2 = .55$$
$$^rM_1M^2 = .53$$

The data indicate that identical genes operate on occluding teeth, causing a strong size correlation between them. The functional necessity for this state of matter is obvious. The surface of the molar forms an intricate pattern of cusps and basins, demanding a correspondingly modeled surface to occlude with; this is especially vital in bunodont forms with slow wear of the dentition. A constant ratio between the dimensions of the occluding teeth is hence necessary, or else the patterns will be at variance with each other, and the function of the teeth is disturbed.

Lewentz and Whiteley (1902) found a somewhat similar characteristic of the correlation in the human hand. The highest correlation was found between lateral and not between longitudinal neighbors, each bone "being on an average more nearly related to the corresponding bone on the next digit, than to the adjacent bone on the same digit." This may be a result of the order of organization of the tissues, which proceeds from the proximal to the distal parts during early development.

TOPOGRAPHY OF THE CORRELATION FIELD: PHALANGER URSINUS

The next series of teeth to be considered is somewhat more extensive, though confined to the lower dentition: $P_4 - M_4$ in *Phalanger ursinus* Temminck. The data are from Hooijer (1952b). The sample (48 specimens) is not fully homogeneous, since it contains equal numbers of the two subspecies, *Ph. u. ursinus* and *Ph. u. togianus* Tate. These have been considered separately as well as in combination; in the latter case, formula (i) was used (see table 1.4).

P_4–M_4 are all subequal in size in *Ph. ursinus,* and form a functional unit. Within this unit the rule of neighborhood holds good to some extent, but not always; moreover, there are differences between the samples. Thus the high point of correlation is between M_2 and M_4 in *Ph. u. ursinus,* and between M_1 and M_2 in *Ph. u. togianus.* P_4–M_1 and M_{3-4} do not show significant correlation in the latter subspecies. Analysis of the significance of these differences (by converting *r* into Fisher's *z* and calculating *d*/σ*d*), however, showed that only rM_1M_4 and rM_2M_4 differed significantly (*P* about .05 in the former, and .01 in the latter case). The other differences are minor, and it is indeed probable that the combined sample gives a better representation of the actual correlation fields than either one of the subsamples; it seems however likely that the subspecies are distinguished by slight modifications of this pattern, caused by genetic differentiation.

Table 1.4
Coronal Length Correlation between P_4—M_4 in *Phalanger u. ursinus* Temminck and *Ph. u. togianus* Tate, and Combined for Both, N = 24 in Both Subsamples. Raw Data from Hooijer. Table of *r*

Phalanger u. ursinus

M_1	M_2	M_3	M_4	
.49	(.38)	(.17)	.45	P_4
	.50	(.22)	(.38)	M_1
		.47	.68	M_2
			.52	M_3

Phalanger u. togianus

M_1	M_2	M_3	M_4	
(.26)	.46	(.27)	(.35)	P_4
	.74	(.33)	(−.17)	M_1
		.48	(.06)	M_2
			(.26)	M_3

Phalanger ursinus (combined)

M_1	M_2	M_3	M_4	
.40	.41	(.25)	.44	P_4
	.61	(.27)	(.18)	M_1
		.47	.45	M_2
			.40	M_3

The point for point study of tables like 1.4 is a rather tedious undertaking, and the main lines tend to drown in the mass of detail. The properties of the correlation fields were found to be brought forth very clearly, if the table was taken as a basis for a topographic diagram. To this end, r at first should be transformed into Fisher's z (which is anyway necessary if comparisons are to be made); the use of z gives a better view of the topography of the field, especially with high correlations, as the distribution of z is approximately normal, whereas that of r is not. The dentitions are marked along two of the borders of the graph, the arrangement being similar to that in the table. The values of z are entered at the points of intersection of the coordinates from the midpoints of the teeth, and after that, contour lines for various values of z can be drawn. In the present case ($N = 24$ for each subsample), $P < .05$ for $z > .40$, and the areas of $z < .40$ have been filled in with black. The correlation fields will then be represented as "islands", arising from a "sea" of noncorrelation or low correlation. The contour lines represent $z = .50$, $z = .60$, etc. In the graph (fig. 1.2) the actual values of z have been left out, in order to avoid unnecessary cluttering of the picture; instead, plus and minus signs indicate peaks and troughs.

Graphs of this kind were found not only to convey a much clearer picture of the conditions under study, but also to provide an excellent basis for visual comparison between different samples. The most striking fact is perhaps that the distribution of high and low correlation values is not random, but of great regularity, so that the contour lines can be drawn smoothly without any special effort, and form characteristic and recurrent

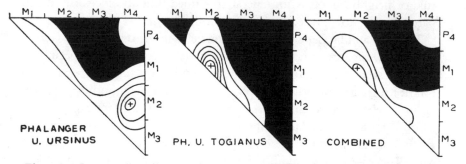

Fig. 1.2   Contour line diagram showing topography of correlation in the series P_4—M_4 in *Phalanger u. ursinus* and *Ph. u. togianus*. Strength of correlation expressed in Fisher's z. Black areas indicate z < .40; interval between lines .10.

patterns in the graph. The diagonal upper left–lower right represents the correlation between adjacent teeth, and in accordance with the rule of neighborhood the highest values of z tend to concentrate along this diagonal, which thus stands out as a ridge in the graph. The black field pictures low correlation, or absence of correlation, between teeth further removed from each other. The "island" in the upper right corner represents significant though low correlation between the end members (in *Ph. u. ursinus* and the combined sample; in *Ph. u. togianus* the value is but slightly less than $z = .40$). The graph is thus a direct "translation" of the table, and comparison is recommended if the graph is not directly understood. The subsequent discussion will to a large extent be based on contour line diagrams of this type.

MISCELLANIA

The samples treated so far have been discussed in some detail in order to draw attention to various properties of the correlation fields and to demonstrate fully the methods employed. In the present paragraph subsection some other data, mostly on "primitive" dentitions—series without diastemata or vestigial teeth—will be briefly considered.

Table 1.5 and fig. 1.3 demonstrate the correlation fields in the entire dentition of the orangutan *Pongo p. pygmaeus* on data from Hooijer (1948). Both sexes are combined by use of formula (i). The sample was fairly large, amounting to 55 skulls, but the topography is somewhat irregular. The rule of neighborhood is in general confirmed, but the incisors, and in particular I^1, show low correlation with other teeth.

Table 1.6 and fig. 1.4 show correlation fields in the dentition of *Tapirus i. indicus* Desmarest; the sample (Hooijer 1947a) comprises 25 specimens, sexes not separated. The small size of the sample is presumably responsible for the irregularity of the topography, but the rule of neighborhood appears again to be confirmed, and there appears also a tendency for raised correlation between end members of the series. The table also lists correlation coefficients for the milk premolars, unfortunately on all too insufficient data. The rule of neighborhood is shown to apply, but relatively low correlation between milk teeth and molars is indicated. Study of the correlation between milk teeth and succeeding permanent teeth would be of singular interest and might throw light on the question whether identical

Table 1.5
Correlation between Crown Dimensions (ap, anteroposterior; tr, transverse) in Orang-Utan Dentition (*Pongo pygmaeus pygmaeus*, 55 Skulls, Males and Femaled Combined, N from 15 to 53. Raw Data from Hookjer.
Table of *r*

| I^2 | C^S | P^3 | P^4 | M^1 | M^2 | M^3 | | |
ap	tr	tr	tr	tr	tr	tr		
(.40)	(.37)	(.38)	(.38)	(.42)	.50	(.41)	I^1	ap
	.75	.70	.47	.55	.50	(.27)	I^2	ap
		.70	.65	.56	.48	(.39)	C^S	tr
			.74	.64	.53	.49	P^3	tr
				.78	.80	.70	P^4	tr
					.86	.79	M^1	tr
						.86	M^2	tr

| I_2 | Ci | P_3 | P_4 | M_1 | M_2 | M_3 | | |
ap	tr	tr	tr	tr	tr	tr		
.84	.66	.77	.74	.69	.67	.57	I_1	ap
	.67	.78	.70	.63	.67	.52	I_2	ap
		.71	.66	.56	.69	.68	Ci	tr
			.82	.58	.58	.38	P_3	tr
				.58	.77	.48	P_4	tr
					.91	.74	M_1	tr
						.81	M_2	tr

| I_1 | I_2 | Ci | P_3 | P_4 | M_1 | M_2 | M_3 | | |
ap	ap	tr	tr	tr	tr	tr	tr		
(.47)	.58	(.34)	(.40)	(.18)	(.41)	(.36)	(.17)	I^1	ap
.77	.80	.70	.86	.64	.64	.68	.59	I^2	ap
.62	.63	.84	.75	.67	.54	.67	.64	C^S	tr
.59	.54	.73	.72	.70	.64	.65	.60	P^3	tr
.59	.57	.59	.65	.82	.71	.74	.61	P^4	tr
.65	.62	.60	.62	.60	.85	.83	.68	M^1	tr
.52	.52	.50	.41	.55	.76	.83	.74	M^2	tr
(.33)	(.41)	.54	(.30)	(.45)	.71	.74	.78	M^3	tr

genes do or do not act on the two tooth generations, but the acquisition of raw data would be very difficult, living material being required.

As an instance of surprisingly low correlation between adjacent teeth may be noted P_4–M_1 in *Smilodon californicus* Bovard (data from Merriam and Stock 1932); the sample (25 mandibles) is certainly heterogenous sexually, but r = .62 only.

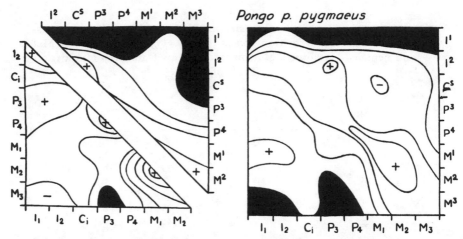

Fig. 1.3.   Contour line diagrams showing correlation fields within upper and lower dentitions (left) and between upper and lower dentitions (right) in *Pongo p. pygmaeus*. Black areas, $z < .50$; interval between lines .20.

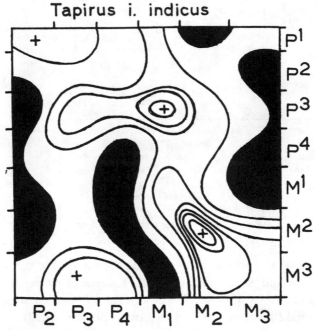

Fig. 1.4.   Correlation fields in dentition of *Tapirus indicus*. For explanation see fig. 1.3.

Table 1.6
Coronal Length Correlation between Cheek Teeth of Recent Indian Tapir, *Tapirus i. indicus* Desmarest, Both Sexes Combined, from Sample of 25 Specimens; N from 5 to 18 Raw Data from H ijer. Table of *r*

Upper tooth row

P^2	P^3	P^4	M^1	M^2	M^3	
(.38)	(.45)	(.50)	.70	(.46)	.71	P^1
	.51	(.40)	.60	(.45)	(.53)	P^2
		.77	.76	(.49)	(.53)	P^3
			.77	.71	(.45)	P^4
				.67	(.37)	M^1
					.92	M^2

Both tooth rows

P	P^2	P^3	P	M^1	M^2	M^3	
.65	.52	(.40)	.59	(.45)	(.28)	.59	P_2
.64	.55	.76	.60	.57	(.52)	.80	P_3
.56	(.48)	.74	(.45)	(.31)	(.33)	.73	P_4
.64	.64	.87	.61	.80	.61	(.37)	M_1
(.45)	.59	.62	.66	.60	.94	.87	M_2
(.58)	(.16)	(.50)	(.33)	(.30)	.76	.85	M_3

Upper Milk teeth and first molar

DP^2	DP^3	DP^4	M^1	
.58	(.40)	(.42)	(.29)	DP^1
	.75	.90		DP^2
		.79		DP^3
			.56	DP^4

CORRELATION FIELDS IN THE DENTITIONS OF THREE CARNIVORES

Vulpes vulpes L. The red fox sample consisted of 20 male and 18 female skulls at the Zoological Institute, Helsingfors. The results were combined according to Formula (i). As some of the specimens were from the island of Kökar, the majority, on the other hand, being from the Finnish mainland, it was thought possible that stunting in the insular population would necessitate further splitting of the sample. Comparison between the means for insular and mainland specimens, however, did not show any significant difference (as a matter of fact some variates averaged slightly larger in the Kökar foxes). Both left and right teeth were measured on each specimen, and the individual means taken as a basis for the computation of the correlation coefficients.

The correlation fields may be evaluated from table 1.7 and fig. 1.5. They evidently differ in some respects from those described previously. The differences appear wherever the size and, presumably, the functional importance of a tooth is reduced. Thus P 1/1 are barely significantly correlated with each other, P^1 furthermore with P^2, P 4/4 and M 1/1, but not with

Table 1.7
Correlation between Crown Dimensions in Dentition of Red Fox (Vulpes vulpes L.), 32 Skulls, Males and Females Combined, N from 30 to 32. Original Data. Table of r

P^1	P^2	P^3	P^4	M^1	M^2	
(.35)	.66	.81	.74	.67	(.31)	C^S
	.53	.41	.66	.57	(.38)	P^1
		.80	.71	.55	.56	P^2
			.65	.52	(.39)	P^3
				.78	.55	P^4
					.48	M^1

P_1	P_2	P_3	P_4	M_1	M_2	M_3	
(.36)	.56	.70	.60	.66	(.30)	(.29)	C_i
	(.25)	(.19)	(.24)	(.39)	(−.03)	(−.10)	P_1
		.76	.65	.57	.45	(.38)	P_2
			.81	.63	.50	(.31)	P_4
				.68	.52	(.38)	P_4
					.62	(.17)	M_1
						(.32)	M_2

C_i	P_1	P_2	P_3	P_4	M_1	M_2	M_3	
.90	(.32)	.55	.67	.63	.69	(.40)	(.26)	C^S
(.29)	.57	(.35)	.43	.49	.58	(.17)	(−.11)	P^1
.62	(.37)	.80	.87	.74	.71	.59	(.29)	P^2
.75	(.38)	.70	.82	.80	.61	.48	(.33)	P^3
.63	.42	.55	.64	.67	.82	.49	(.21)	P^4
.59	(.39)	.44	.46	.54	.86	.54	(.12)	M^1
(.10)	(.08)	(.23)	.46	.48	.43	.60	(−.12)	M^2

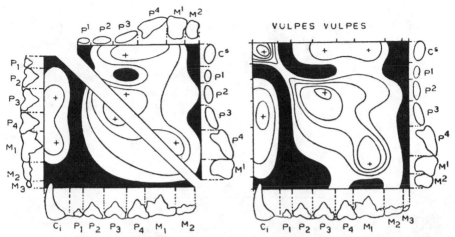

Fig. 1.5. Correlation fields in dentition of *Vulpes vulpes*. for explanations see fig. 1.3.

any other teeth; and P_1 is not significantly correlated with any other tooth than P^1. M_3, most reduced of all teeth in the battery, varies independently of all other teeth. M^2, reduced but not vestigial, shows weak correlation with close neighbors, and none with teeth at a greater distance. Thus, in the contour line graphs, there appears a cross-formed depression, reflecting low correlation values for P 1/1, and marginal lows, indicating the same for M 2/3.

Otherwise, in the premolar-molar series, the rule of neighborhood holds good, and the slight deviations may be due to small sample effects. The canines however show distinct and highly significant correlation with members of the cheek tooth series. Finally, the correlation between occluding teeth is very high, especially that between upper and lower canines.

The picture is thus decidedly more complicated than in the previously described cases, and the presence of reduced or vestigial teeth is seen to result in very steep correlation gradients and fragmentation of the fields.

Ursus arctos L. The fragmentation is even more marked in the bear (table 1.8, fig. 1.6). The sample consists of 16 males and 13 females from Finland (at the Zoological Institute, Helsingfors), and 4 males and 5 females from Sweden (at the Zoological Institute, Uppsala). The Swedish and Finnish populations did not show any significant size differences, but males and females were considered separately, as before. Mensuration and computa-

Table 1.8
Correlation between Crown Lengths in Dentition of Brown Bear (*Ursus arctos L.*), 38 Skulls, Males and Females Combined, N from 32 to 38. Original Data. Table of r

(P^1-P^2)	P^3	P^4	M^1	M^2	
–	(.25)	(.37)	.46	(.33)	C^s
	–	–	–	–	(P^1-P^2)
		.44	.50	.42	P^3
			.61	.64	P^4
				.64	M^1

(P_1-P_3)	P_4	M_1	M_2	M_3	
–	(.41)	.49	(.36)	.44	Ci
	–	–	–	–	(P_1-P_3)
		.78	.65	.46	P_4
			.88	.55	M_1
				(.39)	M_2

Ci	(P_1-P_3)	P_4	M_1	M_2	M_3	
.91	–	.42	.42	(.23)	.49	Cs
–	–	–	–	–	–	(P_1-P_2)
–	–	.68	.49	.40	(.37)	P_3
(.38)	–	.60	.64	.67	.40	P_4
.56	–	.59	.82	.80	(.39)	M^1
(.33)	–	.48	.47	.61	.63	M^2

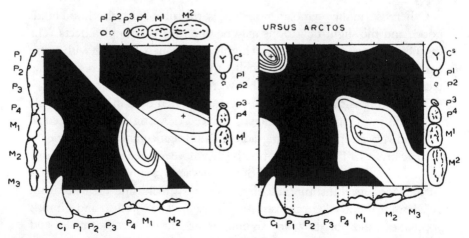

Fig. 1.6. Correlation fields in dentition of *Ursus arctos*. For explanations, see fig. 1.3.

tion of the correlation coefficients was done in the same way as regarding the foxes.

The correlation fields of the bear show gradients resembling those in the fox, but on an exaggerated scale as regarding the premolars, evidently due to the great reduction of these. The cross-formed depression, accordingly, is much widened. P $1-3/1-3$ are vestigial, and of these, P^3 only shows significant correlation with its closest neighbors. In the hind part of the cheek dentition, the rule of neighborhood is confirmed as usual; the apparent decrease in correlation between M^{1-2} and M_{2-3} respectively may be spurious. The canines seem to be less strongly correlated with the cheek teeth than in the fox.

Felis rufus Schreb. The data on the American bobcat are from Merriam and Stock (1932). This sample was very heterogeneous, comprising three subspecies with both sexes represented. In this case the means were computed separately for each of the six subsamples, and the correlation coefficients determined by means of an extension of formula (i). The procedure was checked by a similar study on a smaller sample of *Felis lynx* L. at the Zoological Institute, Helsingfors. No significant differences were found. The data on *F. rufus* do not include the M^1; the coefficients for this tooth pertain to the Finnish *F. lynx* sample. The results (table 1.9, fig. 1.7) are thus compound, but it seems probable that they have some general validity for the subgenus *Felis (Lynx)* at least.

With the complete elimination of the anterior premolars, the cross-formed depression has disappeared except for some vestiges, and the diagonal ridge indicating high correlation between occluding and neighboring teeth is unbroken. A remarkable feature is the fairly high correlation between the vestigial M^1 and its neighbors; whether this may reflect a remaining functional importance or may be spurious is uncertain. The canines, again, show somewhat raised correlation with the cheek teeth, and

Table 1.9

Correlation between Coronal Dimensions in Dentition of Bobcat, *Felis rufus* Schreb., 23 Skulls, Males and Females of Three Subspecies Combined; Data from Merriam and Stock. Data on M_1 from *Felis lynx* L. Table of r

C_i	P_3	P_4	M_1	
.84	.49	.63	.56	C^s
(.43)	.69	.68	(.45)	P^3
.56	(.44)	.74	.73	P
–	–	–	.66	M^1

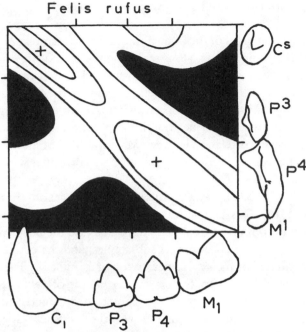

Fig. 1.7.   Correlation fields in dentition of *Felis rufus* (data on M^1 supplemented from *F. lynx*). For explanations see fig. 1.3.

particularly, as in the two other carnivores, with the medial part of the cheek series. High points of correlation are found between the upper and lower canines and in the medial part of the functional cheek series, features parallelled by the foxes and the bears.

CORRELATION FIELD IN THE WING QUILLS OF COLUMBA LIVIA, AND IN OTHER SERIAL STRUCTURES

I have extended the study to a quite different sort of serial structures, viz., the primaries of a bird, to investigate whether this sort of field action occurs in other cases than mammalian dentitions. The species selected was the domestic pigeon, *Columba livia* L. Most of the wings were supplied by Dr. K. O. Donner, to whom I tender my cordial thanks. As all specimens were adults, no growth series is included. The lengths of the ten hand quills and the outermost arm quill were measured. In worn quills the length of the broken tip was approximated; the error was certainly less than 1 mm.; mensuration was carried out with a rule, as recording of decimal places would be neither accurate nor desirable. The number of individuals was 39.

The results appear from table 1.10 and fig. 1.8. The contour line diagram is so closely similar to some of the previous ones that it might almost equally well represent the correlation field of an undifferentiated mammalian dentition. The same general features appear: the diagonal ridge, reflecting the rule of neighborhood; the "low areas" farther from the diagonal, reflecting low or none correlation between members on a greater distance from each other; and the corner "peak," showing ascendant correlation between end members (of the functional unit, the ten primaries).

There are however some additional observations worthy of mention. Thus the highest values of correlation are found in the anterior parts of the series; these are the long primaries that form the aerodynamically important tip of the wing. As their length ratios determine the curvature of the wing point, it is easily seen that high degree of correlation should have selective value. The posterior hand quills are shorter, and their aerodynamic significance, though certainly not small, is probably secondary to that of the anterior ones; and it is seen again that the correlation decreases posteriorly, very regularly.

I fail to find any functional basis for the significantly higher correlation

Table 1.10
Axial Length Correlation between Hand Quills (I . . . X) and First Arm Quill (1) in Domestic Pigeon (*Columba livia*). N = 39. Original Data. Table of r

II	III	IV	V	VI	VII	VIII	IX	X	1	
.91	.76	.68	.76	.65	.64	.53	.47	.57	.73	I
	.91	.83	.86	.77	.71	.55	.52	.50	.58	II
		.97	.88	.78	.64	.53	(.43)	.50	(.42)	III
			.89	.84	.66	.54	(.41)	.50	(.36)	IV
				.84	.63	.51	(.42)	(.45)	(.43)	V
					.78	.61	(.44)	(.45)	(.41)	VI
						.79	.71	.67	(.40)	VII
							.78	.75	.53	VIII
								.84	(.45)	IX
									(.41)	X

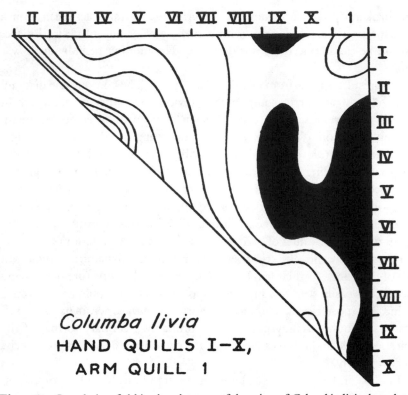

Fig. 1.8. Correlation field in the plumage of the wing of *Columbia livia*, based on axial length of the ten hand quills and the outermost arm quill. For explanations see fig. 1.3.

between the first and last primaries, and also for that between the first primary (I) and the first arm quill (1), features parallelling those of mammalian dentitions, but obscure in meaning. These peculiarities would seem to be conditioned by some kind of inherent mechanics in the development within the field, rather than by adaptive demands. It is to be hoped that future research will contribute to the understanding of the factors involved in this problem.

These characters—the rule of neighborhood and the correlation between end members—emerge clearly if the means of the z's for quills at various distances from each other are calculated. The mean values of z (and corresponding values of r) were found to be as follows:

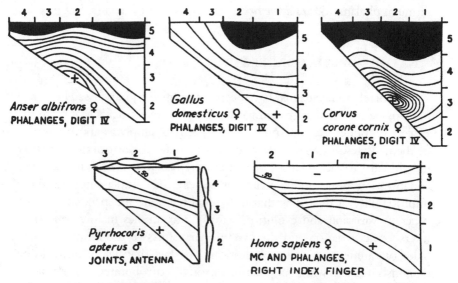

Fig. 1.9. Correlation fields in fourth digit of *Anser Albifrons, Gallus domesticus,* and *Corvus corone cornix* females (length of phalanges): in antenna of *Pyrrhocoris apterus* males (length of joints); and in right index finger of human females (length of metacarpal and phalantes). Black area, z < .40; interval between lines .10.

Distance Between Quills (In Steps)

	1	2	3	4	5	6	7	8	9	10
z	1.26	.95	.80	.67	.56	.56	.52	.49	.65	.93
r	.85	.74	.66	.59	.51	.51	.47	.46	.57	.73

In fig. 1.9 I have brought together some other correlation data in order to present them in a form facilitating comparison. The data on *Homo sapiens* are from Lewentz and Whiteley (1902), those on *Anser albifrons, Gallus domesticus, Corvus corone,* and *Pyrrhocoris apterus* from Alpatov and Boschko-Stepanenko (1928). The rule of neighborhood appears almost throughout, the fourth and fifth phalanges of *Anser albifrons* and *Corvus corone* providing apparent exceptions; moreover the general topography of the fields is closely similar, the gradients in general climb proximally (this is the rule that distal elements show weaker correlation than proximal ones, stated by Alpatov and Boschko-Stepanenko), and the high point occurs between proximal and second, or second and third element. Occasionally the peculiarity of raised correlation between end members (*Homo sapiens,* mc and terminal phalanx; *Gallus domesticus,* phalanges 1 and 5) recurs, but it is not universal.

NATURE OF THE CORRELATION FIELDS

The correlation fields that have been exemplified and discussed here may perhaps be regarded as related to the morphogenetic fields that were the subject of Butler's (1937, 1939) studies; perhaps simply as another aspect of them. Bateson (1892) early pointed out that "the attribution of strict individuality to each member of a series of repeated parts leads to absurdity and in variation such individuality can be set aside." Butler emphasized the improbability of the "view that individual teeth are independent of each other in variation and evolution" (1939). His studies mainly concerned interspecific variation and the evolutionary shifts of the morphogenetic fields of caninization and molarization. The fact that field action can be demonstrated also within populations strongly corroborates Butler's views, and directs attention to the genetic basis of the fields, as well as to the developmental-physiological processes leading to its phenotypic realization.

The presence of such fields is certainly not restricted to instances like those considered here. They bring to one's mind the discovery by Huxley (1932) of growth fields and growth gradients; the demonstrated field action of the determinants of ontogenetic development, and so on; and morever, beyond the scope of biology proper, the increasing importance of the field concept in such disciplines as psychology and physics.

No attempt will here be made to integrate the facts into the general theories of relative growth, but some comparison with Butler's morphogenetic fields is clearly appropriate. Butler (1939) is of the opinion that the morphogenetic field of the mammalian dentition probably has an antero-posterior axis, "for there is no tendency of the posterior end to return to the condition of the anterior end." In the correlation field, on the other hand, there seems frequently to occur a tie between first and last members of a functional unit, suggesting something like a circular "axis" (the term is obviously not very appropriate).

The teeth are not only intercorrelated; they may also have a direct relation to skull size. In *Vulpes vulpes* the functional teeth of adult animals showed a definite and sometimes very high correlation with skull length (table 1.11). Curiously enough, no such correlation appears in *Ursus arctos;* here

Table 1.11
Correlation Between Length of Skull and Coronal Dimensions of Teeth in Red Fox *(Vulpes vulpes* L.), **32 Skulls, Males and Females Combined.**
Table of r

Upper Dentition		Lower Dentition	
C^S	.84	Ci	.78
P^1	(.36)	P_1	(.38)
P^2	.58	P_2	.56
P^3	.76	P_3	.67
P^4	.63	P_4	.64
M^1	.66	M_1	.69
M^2	(.30)	M_2	(.35)
		M_3	(.05)

r fluctuates around zero, with random positive and negative values. The difference between the species is surely significant.

Wright (1932) has discussed the distinction between general, group, and special factors that govern the growth and, *a fortiori,* the ultimate size of organs. It would seem that the mammalian dentition is in general governed by group and special factors; but in certain instances, as in the fox, part of the variation is seemingly due to the general factors conditioning skull and gross size. The possibility that this correlation might result from extrinsic influences would seem to be denied by the difference between *Ursus arctos* and *Vulpes vulpes;* for if environmental conditions impose a phenotypic correlation in the latter species, why not in the former?

Yet part of the variation is certainly nongenetic. It can hardly be doubted that homologous right and left teeth are governed by identical genes. Yet the correlation between them is not absolute; *r* < 1. Table 1.12 shows the correlation between selected right and left teeth in bears and foxes. This

Table 1.12
Correlation Between Left and Right Canines and First Premolars in Red Fox *(Vulpes vulpes* L.) **and Brown Bear** *(Ursus arctos* L.). **Males and Females Combined. Table of r**

Vulpes vulpes		Ursus arctos
Cs	.91	.91
Ci	.96	.84
P^1	.88	.66
P_1	.58	

correlation is invariably high for fully functional teeth, and progressively lowered in reduced and vestigial teeth. The values for functional teeth are of the same order as those for the correlation between functional antagonistic teeth—for instance, the canines in carnivores—and it seems justifiable to assume genetic identity in these latter cases as well.

The differences between correlations in males and females have not been touched upon earlier. It must however be emphasized that such differences occur and that they seem to be systematic, viz., that correlation is higher in the male dentition in all species where analysis has been possible. In order to demonstrate the magnitude of the differences, I have computed separate coefficients for males and females, transformed to z and tabulated the deviation (table 1.13, $z\delta - z\female$). These differences are rarely of real statistical significance, but the evidence is cumulative.

It is perhaps trivial to offer the concluding remark that the correlation pattern of the diagrams naturally is something quite different from the genes that partake in its causation. Its structure does not reveal the structure of the genes; but it does throw some light on the structure of their action. To be sure, it may be influenced by other factors that may confuse some characters and obliterate others, but the general features seem to be valid. As the concept is somewhat difficult, a simile might be employed. When an electronic ray is projected through a metal foil there ensues a pattern of interference rings. This does not show what the electron actually looks like; yet it tells something about how it behaves. In a somewhat similar, though perhaps even more roundabout way, the topography of the correlation field tells something about how the gene behaves. The analogy should however serve as a warning against over-simplified pictures; and to the correlation field, as well as to the morphogenetic, is certainly applicable Butler's pithy conclusion—"The field is probably not simple in structure."

Evolution of the Correlation

In the previous chapter a comparison was made between dental correlation fields on different levels of specialization. The present chapter comprises a study of an actual case of phylogenetic evolution of such growth fields.

The Pontian Chinese Hyaenidae furnish an interesting instance of the evolution of the correlation in connection with the reduction of part of the

Table 1.13
Differences (z♂−z♀) between Transformed Correlation Coefficients for Males and Females in Red Fox, *Vulpes vulpes* L., and Brown bear, *Ursus arctos* L.

Vulpes vulpes

P^1	P^2	P^3	P^4	M^1	M^2	C_i	P_1	P_2	P_3	P_4	M_1	M_2	M_3	
.21	.06	.48	.40	−.41	.24	.48	.32	−.03	.55	.91	.17	.11	.22	C^S
	−.06	.12	−.30	−.27	.18	.17	−.17	−.10	.04	−.03	−.15	−.02	.60	P^1
		.26	−.05	−.29	.08	.20	.15	.08	.55	.59	.12	.32	−.63	P^2
			.32	−.19	.36	.64	.61	−.01	.53	.49	.38	.31	−.23	P^3
				−.50	−.17	.41	.19	.06	.05	.34	−.26	.04	.17	P^4
					.33	−.17	.18	−.25	−.06	.19	−.60	.07	.37	M^1
						.45	.16	.51	.04	.04	.28	.26	.55	M^2
							.27	−.23	.55	.86	.25	.21	.07	C_i
								−.15	.39	.86	−.01	.28	.15	P_1
									.37	.54	.11	.67	−.11	P_2
										.58	.53	.38	.09	P_3
											.57	.21	.35	P_4
												.05	⊣	M_1
													⊣	M_2

Ursus arctos

P^3	P^4	M^1	M^2	C_i	P_4	M_1	M_2	M_3	
.59	.81	.37	.87	.07	.24	.05	.55	.68	C^S
	.49	.83	.40	—	.26	1.06	.63	.27	P^3
		.53	.35	.68	.12	.29	.06	.59	P^4
			1.06	.16	.40	−.04	.40	.23	M^1
				.56	.23	.24	.62	−.38	M^2
					.30	−.14	.38	.55	C_i
						.71	.38	.18	P_4
							−.19	.36	M_1
								.37	M_2

dental battery. Before entering on this subject, however, some inquiry must be made into the status of the hyaenid populations in the Lagrelius collection.

THE HYAENID POPULATIONS

The Hyaenidae of the Chinese *Hipparion* fauna were thoroughly described by Zdansky (1924, 1927), whose treatment revealed a new variation

of the common theme of contemporary populations forming a morphological intergradation between two evolutionary stages. Somewhat similar
series are present in other parts of the *Hipparion* fauna, but seemingly none
more explicit than this one.

On account of the intergradation it is a somewhat delicate task to find
out where one species ends and the next one begins. Only one of them is
well separated from the others, viz., *Crocuta variabilis* Zdansky, sole representative of its genus in the fauna.

This species may be contrasted with the entire group of smaller hyaenids
(*Ictitherium–Lycyaena*) in the collection. Tables 1.14 and 1.15 illustrate the
frequency distributions of two variates. If the variation within the *Crocuta*
species is considered to be typical of a homogeneous hyaenid population,

Table 1.14
Length of P⁴ in Pontain Hyaenidae of China

	Frequencies				
	Ictitherium				
Class Midpoint	gaudryi	wongii	hyaenoides	Crocuta variabilis	Total
18.95	2	–	–	–	2
19.95	4	–	–	–	4
20.95	2	–	–	–	2
21.95	1	–	–	–	1
22.95	–	1	–	–	1
23.95	–	9	–	–	9
24.95	–	11	–	–	11
25.95	–	9	–	–	9
26.95	–	6	5	–	11
27.95	–	–	5	–	5
28.95	–	–	2	–	2
29.95	–	–	2	–	2
30.95	–	–	2	–	2
31.95	–	–	–	–	(2)[1]
32.95	–	–	–	–	–
33.95	–	–	–	–	–
34.95	–	–	–	1	1
35.95	–	–	–	1	1
36.95	–	–	–	4	4
37.95	–	–	–	4	4
38.95	–	–	–	1	1
39.95	–	–	–	1	1

Table 1.15
Length C–P⁴ in Pontian hyaenidae of China

	Frequencies				
	Ictitherium				
Class Midpoint	gaudryi	wongii	hyaenoides	Crocuta variabilis	Total
56.5	1	–	–	–	1
60.5	–	–	–	–	–
64.5	1	–	–	–	1
68.5	3	–	–	–	3
72.5	1	3	–	–	4
76.5	–	4	1	–	5
80.5	–	14	–	–	14
84.5	–	2	2	–	4
88.5	–	–	2	–	2
92.5	–	–	2	–	2
96.5	–	–	–	–	–
100.5	–	–	–	1	1
104.5	–	–	–	4	4
108.5	–	–	–	3	3

it may be tentatively assumed that the variation in related populations should be of the same order of magnitude.[1] If so, the *Ictitherium–Lycyaena* group is clearly not honogeneous. The observed range and the modes indicate that the distributions within the group are really compounds of three samples, differing as to their means, and conforming to Zdansky's three main species, *I. gaudryi, I. wongii,* and *I. hyaenoides.*

That this is really so can be easily demonstrated from a graph combining the two distinctive variates, length of P^4 and breadth of M^2, in a scatter diagram (fig. 1.10). The picture is clearly that of three main groups, though some borderline cases and aberrant specimens occur.

The small *I. gaudryi* is fairly well isolated; it is the structural ancestor of the series, and has a large M^2. By far the most numerous population in the collection is that of *I. wongii.* This form is readily separable from *I. gaudryi,* being larger and having a smaller M^2. It is not so easily distinguished from the (on an average) still somewhat larger *I. hyaenoides;* there are several borderline cases, and when the material is damaged so that the M^2 is lacking, the proper status is often impossible to determine.

There are several unclassified specimens, not treated before, the majority

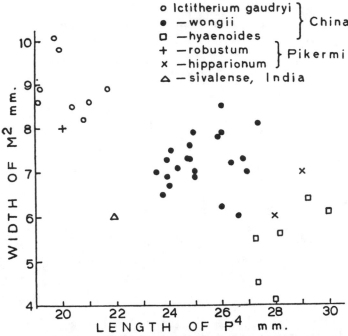

Fig. 1.10.   Scatter diagram showing breadth of M^2 plotted against length of P^4 in Pontian Hyaenidae from China, as labelled. For purposes of comparison, data on Hyaenidae from Pikermi (after Gaudry) and India (after Pilgrim) are recorded.

of which may unhesitatingly be referred to *I. wongii*. When these were included in the variation analysis, the means remained practically constant, as did the standard deviations. Some parameters are given in table 1.16.

There is indeed some probability for the presence of precisely three congeneric populations, for three is some evidence that there were three main biotopes—steppe, forest, and an intermediate biotope in between (see Kurtén 1952). The three species of gazelles, *Gazella dorcadoides, G. gaudryi,* and *G. paotehensis,* seem to be distributed in a similar fashion; among the sabre-tooths, at least one species *(Paramachairodus maximiliani)* occurs in the forest zone of the south, while two larger forms, *Machairodus palanderi* and *M. tingii,* have been found further northward.

Table 1.16
Variation in Chinese pontian Hyaenidae. Original Data.

	N	M	σ	V	S. R. (S. D.)
Ictitherium gaudryi					
Length P⁴	9	20.1 ± .27	.82 ± .19	4.1 ± 1.0	17.4 ± 22.8
Breadth M¹	9	14.7 ± .34	1.01 ± .24	6.9 ± 1.6	11.4 ± 18.0
Breadth M²	9	9.0 ± .20	.59 ± .14	6.5 ± 1.5	7.1 ± 10.9
Ictitherium wongii					
Length P⁴	36	25.3 ± .19	1.11 ± .13	4.4 ± .5	21.7 ± 28.9
Breadth M¹	28	15.2 ± .19	1.00 ± .13	6.6 ± .9	12.0 ± 18.4
Breadth M²	28	7.4 ± .15	.78 ± .10	10.4 ± 1.4	4.9 ± 9.9
Ictitherium hyaenoides					
Length P⁴	18	28.7 ± .4	1.66 ± .28	5.8 ± 1.0	23.3 ± 34.1
Breadth M¹	17	15.9 ± .34	1.40 ± .24	8.8 ± 1.5	11.4 ± 20.4
Breadth M²	7	5.4 ± .28	.73 ± .20	13.6 ± 3.6	2.9 ± 7.9
Crocuta variabilis					
Length P⁴	11	37.5 ± .41	1.36 ± .29	3.6 ± .8	33.1 ± 41.9
Breadth M¹	10	14.5 ± .32	1.02 ± .23	7.1 ± 1.6	11.2 ± 17.8

N, size of sample; M, mean with standard error; σ, standard deviation with standard error; V, coefficient of variation with standard error; S. R. (S. D.), standard range from standard deviation (see Simpson 1941b), estimating variation limits in a sample of 1,000.

ABERRANT SPECIMENS AND VARIATION OF VESTIGIAL TEETH

Beside the four hyaenids discussed above, Zdansky has described one skull under the name of *I. sinense,* and another was doubtfully referred to a new species also, *?Lycyaena dubia.* Further the collection includes one specimen labeled *?Lycyaena sp.* The two former ones will be considered in this section.

In my opinion *I. sinense* may probably be included under *I. gaudryi.* The single specimen was found in a fossil pocket together with some *I. gaudryi* specimens—evidence for temporal and spatial association (see below). M² is large but well inside the probable range of *I. gaudryi.* The aberration of the P⁴, described by Zdansky (1924), may be an individual mutation; to my knowledge it does not recur in other ictitheres, and it would hardly warrant the assumption of a fourth, specifically isolated, *Ictitherium* population within the fauna. This possibility is also pointed out by Zdansky.

The single *?Lycyaena dubia* specimen I am inclined to refer to *I. hyaenoides*. Except for a slightly different orientation of the P^2, apparent from Zdansky's (1924, Taf. XXXIII) figure, hardly of specific value, and the absence of M^2, the specimen appears to be a large *I. hyaenoides*. In this population M^2 is very variable and much reduced ($V = 13.6$, $M = 5.4$), and it may well be absent in some individuals.

A similar case was studied in the brown bear, *Ursus arctos* L. $P^2/2$—3 are more or less vestigial, and any one of them may be missing. The mean size of P_3 has sunk only slightly below phenotypic realization size. In 82 jaw halves, the tooth was present (emerged) in 25 cases, and in 7 more there was an alveolus, the tooth having been lost, perhaps during preparation; the remaining 50 jaw halves did not show any trace of this tooth. The largest diameter of the tooth crown was more than $3\frac{1}{2}$ mm in all cases except two, in which it was 2.9 mm; these teeth were not fully emerged. It may be assumed that these were exceptional cases and that the phenotypic realization threshold of this tooth is at about 3.5 mm. Smaller teeth may probably be formed in the jaw, but not emerge, as a rule. There were, then, 30 teeth above, and 52 below the phenotypic realization threshold.

Some 36 percent of the teeth would thus be above the threshold, which in terms of a normal distribution would place this threshold at about $M + .35$ σ. The present teeth, indeed, fit well into the right-hand part of a normal distribution, the largest number falling between 3.5 and 4.5 mm, and the histogram sloping to the right. The teeth represented by alveoli, and the missing ones, may be arranged (above and below the threshold, respectively) into a histogram with the mean at about 3.0 mm. (fig. 1.11). σ as required. This histogram is probably a fair enough representation of the "genotypic size" distribution of this tooth.

It may be objected that part of the range is brought below zero, which would imply that the tooth were genotypically lost in some individuals. The fact that teeth which are phylogenetically lost may occasionally be phenotypically realized (e.g. M^4 in the orang-utan, see Hooijer 1948) seems to indicate that complete genotypic loss is at least a very slow process. The distribution of *U. arctos* P_3 ought perhaps to be represented on a logarithmic abscissa instead, in which case the arrangement might be as in table 1.17.

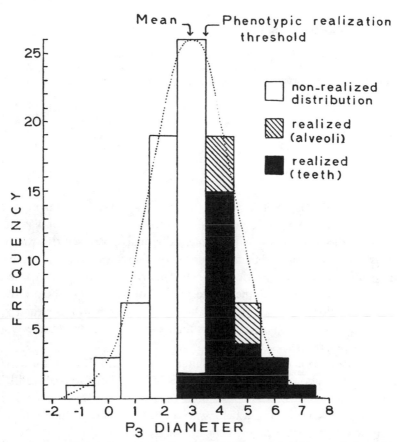

Fig. 1.11. Partly hypothetic distribution of P_3 crown size (largest diameter) in a sample of *U. arctos*.

The standard deviation expressed in log length, = 1.46 and the threshold (log length = .54) is at $M + .34\sigma$.

This distribution is clearly essentially similar to that which must have existed in the first *Lycyaenas* as regarding M^2. It should be emphasized that isolated finds to not permit of certain inferences as to classification, and that two specimens, otherwise of the same type, should not be regarded as drawn from different populations on the evidence of a vestigial tooth only.

This is not said in criticism of earlier authors, nor do I wish to imply

Table 1.17.
Distribution of *U. arctus* P_3 Shown Logarithmically

Length (mm.)	log Length	Frequencies			
		Absent	Present	Alveoli	Total
1.4–1.7	.14–.23	4	–	–	4
1.8–2.1	.24–.33	8	–	–	8
2.2–2.7	.34–.43	18	–	–	18
2.8–3.4	.44–.53	20	2	–	22
3.5–4.3	.54–.63	–	13	5	18
4.4–5.4	.64–.73	–	6	2	8
5.5–6.8	.74–.83	–	4	–	4

that the presence or absence of M^2 in the Hyaenidae lacks taxonomic significance. I merely want to point out that in the transitory populations between *Ictitherium* and *Lycyaena,* some individuals possessed this tooth, whereas others did not, and that, in such borderline cases, single specimens do not suffice for a valid generic reference. The transition may be regarded to have occurred when the mean size of the tooth has sunk below the phenotypic realization threshold and the tooth is absent in more than 50 percent of the population (this may of course have happened within several different populations).

As far as can be deemed from the Chinese sample and the literature, the threshold was at about 3½–4 mm breadth of M^2. The "genotypic breadth" of this tooth in the ?*L. dubia* specimen may then have been anywhere between 0 and 4 mm; if it did belong to the *I. hyaenoides* population, the magnitude was probably closely beneath the threshold value.

As to the other variates, no conclusive proof of heterogeneity can be obtained. Zdansky (1924) indeed thinks it possible "dass es sich um einen Schädel dieser Gattung handelt, an dem als Abnormität M^2 nicht zur Ausbildung gelangt ist." If my interpretation is correct, the word Abnormität is slightly inappropriate, for it would then be simply a distal variant within a normal distribution pattern.

OUTLINE OF HYAENID EVOLUTION

The present consensus of workers in this field is that the Hyaenidae arose from the viverrid stock at some time in the Miocene (Colbert 1935, 1939; Pilgrim 1931, 1932); Gregory and Hellman (1939) consider *Viverra* to be the (structural?) ancestor and the Protelinae to be an early offshoot from the hyaenid branch. The hyaenid family "would seem to have become established very suddenly . . . and to have continued, for the most part without change since then" (Colbert 1939). This is probably an instance of a "quantum shift" (Simpson 1944) from the ecological station, or "adaptive zone" (ibid.) of the viverrids to that of the hyaenids; or, in other words, a carnivore population slumped into carrion-feeding. Preadaptation is not in this case difficult to imagine: carrion-feeding is among the prospective functions (Parr 1926) of practically any predator. The hyaenids were probably scavengers at the *Ictitherium* stage already (Abel 1927); I have elsewhere (Kurtén 1952) suggested that the smaller hyaenids may have been ecological predecessors of the recent jackals.

The actual phase of quantum evolution is not recorded. This is one of the "systematic deficiencies" of the record (Simpson 1944). However, the phyletic or "anagenetic" (Rensch 1947) phase of evolution subsequent to the quantum shift is similarly unrecorded, in contrast with, for instance, the horse sequence. The hyaenids emerge in the record with a number of forms, among which one of the oldest is the very large and highly specialized *Crocuta tungurensis* Colbert. It is evident that the hyaenid "Virenzperiod" (Rensch 1947) was passed at this time, and then, for quite a long time, primitive and more highly specialized hyaenas existed side by side.

One is tempted to localize the unrecorded parts of the evolution of a group in some remote corner of the world that has not been adequately investigated, or else in some area where fossils have not been deposited. This is to think in terms of evolution in situ—in temporal clines, or, geologically speaking, vertical evolution. Kaufmann (1933) has shown, however, that the fossil record often indicates horizontal rather than vertical evolution: spatial clines (Huxley 1939) or Rassenkreise (Rensch 1929). The well-known fact that successive geological horizons from one locality often exhibit what looks like saltatory changes is interpreted as a result of "eine gerichtete Rassenkreisverschiebung" (Kaufmann 1933).[2]

Assume, at one temporal level, a geographic variation from population (a) to population (d) (fig. 1.12), as a result of previous migration in the given direction, during which the border populations of the spreading form have undergone repeated speciation; this is a state of matter abundantly exemplified in neosystematics. Back-migration of population (d), perhaps by a roundabout route, avoiding contact with and genetic swamp-

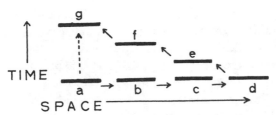

Fig. 1.12. *Stufenreihe* (populations a, g) in successive strata at one locality, arising from geographic speciation and backmigration of terminal population.

ing by related populations (c) and (b), may lead through a new succession to, say, population (g), specifically different from (a). If differentiation has not proceeded very far, competition will probably ensue, and one of the species may eventually become extinct (see Gause et al. 1934). In other cases both species may coexist, occupying different niches. The latter seems to have been true for many fossil hyaenids.

Under such circumstances the branching within the family cannot have been restricted to some isolated area; it clearly requires large territories for repeated migration. The larger and more varied the territory available, the more differentiated the result—in principle.

There is however no fossil record of this phase. Simpson (1936) has shown that intensive sampling from a limited area will generally give a fairly good picture of the basic taxonomic variety of a continental fauna, and that negative evidence (absence) thus may carry some weight. There is, however, in the first place no evidence that we certainly have fossiliferous deposits from the precise period when the hyaenas arose, and, in the second place, no evidence that the sampling in that period was intensive.

One important feature of hyaenid evolution is size increase. This will be more fully dealt with in a later section; suffice it here to give one indication of how hyaenas of different size may have arisen and become sympatric. Supposing (a) (fig. 1.12) to be a small forest form and the migration to proceed northward, (b) might be a savanna form, (c) a steppe form, and (d) a tundra form, with successive size increase in accordance with Bergmann's (1847) rule[3], while (a) remains conservative. When the large size of (d) is attained, a new ecological niche may be available, and the large form then spread back; it may then invade the territory of (a) not as a competitor, but in a different niche.

Such differentiation may, of course, take place in some period of diastrophic change. In land animals a period of mountain-building would seem to be accompanied by rapid speciation. For instance, during part of the Oligocene, the present-day Alpine range was an archipelago (see e.g. Wills 1951); the analogy with avian speciation in the southwest Pacific archipelago (Mayr 1944) is obvious. However, in a recent symposium on this time-honored hypothesis, Simpson (1952) and Newell (1952) do not find any evidence for such correlations between diastrophic and evolutionary data. Their investigation, however, was concerned with world-wide revolutions and global faunal characteristics; local tectonic events, though details in the main picture, may still have important evolutionary results, and evidently do have such at present. The failing to find a general correlation may be connected with the possibility that, as Henbest (1952) hints, the concept of a worldwide diastrophic rhythm may be oversimplified and "the theorizing on rhythm in nature has outrun the facts and possibly its actual importance."

Returning now to the morphological series of Pontian Hyaenidae under consideration, it may safely be stated that perfect contemporaneity is established, since members of every population have been found associated (see below). At Loc. 49, where there seems to have been an extensive mixing of faunal elements from adjoining areas, were found *I. gaudryi, I. "sinense," I. wongii, I. hyaenoides, "?L. dubia,"* and *C. variabilis,* or the complete assemblage; Loc. 109 contains representatives of the four populations and only *I. gaudryi* is lacking at Locs. 30, 43, 108, and 115.

Yet it seems almost axiomatic that the smaller forms represent persisting and relatively little modified ancestors of the larger ones. Recurrent speciation by dichotomy or more excessive fragmentation is the probable explanation of the differentiation. Discussion of phylogeny on the basis of contemporary *Stufenreihen* is liable to serious error (Simpson 1950), but the representation of the reduction of the postcarnassial dentition, as seen in this series, is probably fairly similar to the actual course of evolution; comparable stages are found in Hyaenidae from other territories. The ratio diagram,[4] fig. 1.13, shows that the dimensions in question evolved somewhat in unison, though probably with some irregularities.

EVOLUTION OF CORRELATION FIELDS

The sequence from the stage of *Ictitherium* to that of the true hyaena is characterized by phyletic growth; by increasing robustness of the cheek teeth; and by the dwindling and partial loss of the postcarnassial teeth.

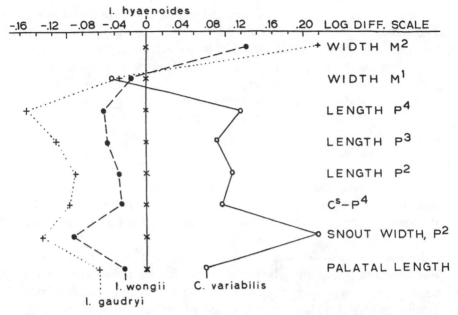

Fig. 1.13   Ratio diagram comparing proportions in upper jaws of Pontian Hyaen-idae from China (means), as labelled. Standard (straight line), *Ictitherium hyaenoides;* means for forms with similar proportions will tend to be arranged in a straight line. Greatest relative increase in size appears in width of snout at P^2 and length of carnassial; smallest relative increase in palatal length, length of tooth row (C—P^4), and dimensions of anterior premolars. First molar shows size increase from *I. gaudryi* to *I. hyaenoides* and subsequent decrease to *Crocuta variabilis;* second molar decreases in size from *I. gaudryi* to *I. hyaenoides,* and disappears in *C. variabilis.*

This evolution is accompanied by significant changes in the correlation fields of the dentition (see table 1.18, fig. 1.14).

In *I. gaudryi,* a single field of correlation seems to have extended from P^2 to M^1 (teeth anterior to these have not been considered in the present series). M^2 seems to have been outside of this field, not being correlated with any other tooth. These results are, to be sure, based on a very small number of individuals; yet the correlation between P^4 and M^1 is probably significant ($P < .05$), whereas the evidence indicates that M^2 varied quite independently.

In *I. wongii* the correlation field would seem to have been contracting, the correlation between P^4 and M^1 being weakened. It is as if M^1 were moving out of the general field (or vice versa), but this is not as yet fully

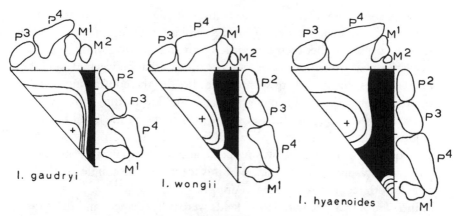

Fig. 1.14. Evolution of correlation fields in upper dentition of ictitheres, as labelled; dentitions to scale. Data on premolars from *I. wongii*. Black area, z < .50. Interval between contour lines .10.

Table 1.18
Changes in Correlation Field of Upper Dentition in the Series, *Ictitherium gaudryi–I. wongii–I. hyaenoides;* Maximum Values of N, 9, 38, and 7, respectively. Original Data. Table of r

	Ictitherium gaudryi			
P^3	P^4	M^1	M^2	
.55	.52	.68	–	P^2
	.73	.77	–	P^3
		.86	–	P^4
			–	M^1

	Ictitherium wongii			
P^3	P^4	M^1	M^2	
.55	.52	–	–	P^2
	.73	–	–	P^3
		.40	–	P^4
			.48	M^1

	Ictitherium hyaenoides			
P^3	P^4	M^1	M^2	
.55	.52	–	–	P^2
	.73	–	–	P^3
		–	–	P^4
			.81	M^1

accomplished, since there remains a statistically significant correlation. On the other hand, there has been formed a significant, though as yet not very strong, correlation between this same tooth and the dwindling M^2. This newly formed correlation was about as strong as that between M^1 and the carnassial. In this way there was formed, besides the contracting general or "premolar field," an incipient "molar field." Up to this point, M^1 increased in size along with the rest of the dentition (fig. 1.14); but from the *I. wongii* stage on, the process was reversed, and, though gross size was subsequently increased, M^1 was reduced, at first relatively, and later on in absolute measure. As the variation of this tooth became coupled with that of M^2, both teeth were reduced in unison.

In *I. hyaenoides*, the correlation fields were fully formed, and the correlation between P^4 and M^1 was eliminated. There was a premolar field, seemingly homologous with the general field in *I. gaudryi*, and this part of the dentition had increased in size along with the gross size of the animal; then there was a molar field, possibly homologous with the incipient one of *I. wongii*, and the teeth within that were in the process of becoming vestigial.

In *Crocuta variabilis*, finally, M^2 had sunk below phenotypic realization size, and M^1 remained as a vestigial tooth, outside of the premolar field.

Along with these changes it can be seen that variation of the reduced teeth was progressively increased (see table 1.16).

It appears that conditions as regarding M^1 in *C. variabilis* were greatly similar to those regarding M^2 in *I. gaudryi*, and the history of M^1 in the series from *I. gaudryi* to *C. variabilis* is probably a repetition of what happened to M^2 before the stage seen in *I. gaudryi*. Moreover, it is probably a valid example of what generally happens when a structure out of a correlated series becomes superfluous and is reduced; the series from an undifferentiated dentition (e.g. the orangutan, fig. 1.3) through the fox (fig. 1.5) to the bear (fig. 1.6) shows a similar process in the progressive detachment of the anterior premolars from the general field. Possibly the superfluous part of the series cannot be effectively reduced as long as it is genetically linked to the functionally important parts adjacent to it: this linkage must break down.

In the present case selection seems to have tended toward reduction of the postcarnassial teeth, possibly because their reduction might bring the carnassials nearer to the jaw joint and thus increase their effectiveness.

The presence, in the *I. wongii* population, of a correlation stage intermediate between those of *I. gaudryi* and *I. hyaenoides* is a fortunate event from the point of view of evolution mechanics. It goes to show that the change did not occur in one saltation, and indicates a cumulative series of small steps, just as in the morphological traits commonly studied in phyletic sequences. The mechanism underlying these processes is, of course, very obscure as yet; but at least some hints as to its structure are given. The probable background would seem to be that the interaction of the genes was delimited slowly, until finally no gene acted on more than one of the two adjacent structures, P^4 and M^1. A similar history may be postulated for the detachment of M^2 from the general growth field, before the *I. gaudryi* stage.[5]

The detachment of a tooth from a correlation field can be easily imitated by means of a model similar to that in table 1.2. The action of the genes may be changed in two steps, somewhat as in table 1.19.

If the teeth, (I)—(III) are assumed to represent P^3—M^1, the stages would correspond to those of *I. gaudryi, I. wongii,* and *I. hyaenoides,* respectively. Thus a relatively slight repression in the range of action of the genes is sufficient to bring about the detachment of a tooth from the general correlation field.

AN ABERRANT SPECIMEN

In addition to the large *"?Lycyaena dubia"* there is also another specimen that lacks the M^2, this is a part of the right upper jaw with a P^4 in size and morphology closely resembling that of *I. wongii,* and an M^1 which is more

Table 1.19.
Model for Detachment of a Tooth from a Correlation Field

	Increment in crown size, caused by gene								
	(First stage)			(Second stage)			(Third stage)		
Tooth No.	A	B	C	A	B	C	A	B	C
(I)	4	2	1	4	2	0	4	2	0
(II)	2	4	2	2	4	1	2	4	0
(III)	1	2	4	0	1	4	0	1	4

correlations change as follows:

	(First stage)	(Second stage)	(Third stage)
r_{I-II}	.83	.75	.85
r_{II-III}	.83	.42	.23
r_{I-III}	.68	.11	.08

reduced than in other members of this population. This might be a true *Lycyaena,* or a small *I. hyaenoides,* or a mutant *I. wongii,,* or a *hybrid form.* The specimen is from Loc. 30₅. Comparison of size by the *t* test gives the result shown in table 1.20.

The size of the premolars excludes this specimen from the range of *I. hyaenoides* and places it within that of *I. wongii.* M^1, however, is too small to suit the pattern of *I. wongii,* and is barely inside the range of *I. hyaenoides.*

There is of course the possibility that this individual was a hybrid, inheriting its gross size from the smaller *I. wongii,* and its small M^1 from *I. hyaenoides;* if the molars were correlated in size, as in *I. hyaenoides,* the small size of M^1 would account for the total disappearance of M^2. Evidence of hybridization from fossils is, however, notoriously uncertain. Schlaikjer (1935) records a possible case in equids *(Mesohippus—"Pediohippus"),* but his conclusions are rejected by Simpson (1944). A well-documented case is that of the human population from Mount Carmel as interpreted by Dobzhansky (1944), but its bearing on the present instance is difficult to evaluate.

Hybridization in mammals under natural conditions is fairly exceptional, but there are some well-documented instances. One of the best analyzed is perhaps that of the two hares, *Lepus europaeus* Pall. and *L. timidus* L., which habitually interbreed wherever their ranges overlap, forming sterile hybrids (Notini 1948). The rut begins somewhat earlier in *L. europaeus,* owing to the southern origin of this species, and the males often breed with *L. timidus* females, whereas the reverse does not seem to take place; in this way *L. europaeus* very quickly attains predominance over its northern ally, and is ousting it, except from forested biotopes.

Such phenomena are however so rare that, if the *?Lycyaena* sp. was really a hybrid, the populations would probably not have been specifically isolated

Table 1.20.
Comparison of Size of Specimen from Loc. 30₅

	Comparison with			
	I. wongii		*I. hyaenoides*	
	t	p	t	p
Length P^3	.4	>.10	2.2	>.10
Length P^4	.5	>.10	2.4	>.05
Width M_1	2.2	>.02	2.1	=.05

from each other. The evidence is however so ambiguous that at present no conclusions, not even tentative ones, can be drawn.

To be sure it would seem an odd picture with two subspecies of the same species having widely overlapping areas of distribution. It must however be borne in mind that the picture revealed by the fossil quarries is really but a haphazard sample from different temporal strata. The distribution within different parts of the territory tends to show that *I. hyaenoides* was the steppe form, *I. gaudryi* the forest form, and *I. wongii* the intermediate. But whereas this is probably true of the general pattern of distribution, the fluctuations may have been much greater than in the gazelles, for example, resulting in the complication of record seen in the mixing of the ictithere populations in the fossil pockets. Presumably the biotopes of the scavenging ictitheres were not so sharply bounded as those of the grazing or browsing gazelles. At each locality, however, one ictithere form generally predominates in numbers, the others being represented by strays; and the true picture may have been one of populations on or slightly beyond the border of specific differentiation, now in marginal contact and in close competition at the overlapping areas, now one and now another gaining the upper hand. Such a history would tend to leave a record confused in details, but with some general features of distribution clearly emerging.

Part II. Population Dynamics

The Meaning of Population Dynamics

Population dynamics is a junior branch of ecology, though its ancestory in the study of human populations extends back almost to the birth of civilization (for historical data, see Pearl 1940). Its basic importance for the study of selection was clearly apprehended decades before the methods were applied to animal ecology (see, e.g., Macdonell, 1913). As Hickey (1952) puts it, the life tables "suggest to the geneticist and the evolutionist the rate to which natural selection may cull out mutations or permit them to become predominant."

The study of population dynamics is as yet in its descriptive stage. As Deevey (1947) points out, to construct life tables for natural populations is not a substitute for the study of the causation of death, but it is an important prerequisite. This latter branch of ecology is progressing well, but it is still a far cry to a general theory of mortality, and there is a great clash of opinions concerning the factors of mortality. This is not the place for a review, and the reader is referred to Solomon's (1949) admirable exposition. It has however become apparent that no single agent can be patted on the back as *the* factor in the natural control of populations (to borrow Simpson's phrase on factors of evolution). The distinction between density-dependent and density-independent factors of mortality (Smith 1935) was a most important step in the study of causes of death. In recent very stimulating and suggestive works, Errington (e.g., 1946) dissents from the orthodox view that predation is the main factor in the stability of natural populations, and stresses the importance of intraspecific competition. The paleontologist, however, familiar with arrays of evolutionary series and trends that can hardly be explained in any other terms than as progressive defensive adaptations, will certainly not be content with any theory that does not consider predation one of the main factors in mortality; but this aspect has possibly been somewhat overemphasized in paleoecological thought, as in early Darwinism (see Simpson's criticism of phrases like "struggle for existence" and "nature red in fang and claw" 1949). To the interpreter of fossil series, knowledge of modern work on population control is indispensable.

The formal population dynamics, crystallized into the life table, is thus but one of a number of related subjects awaiting future synthesis: it will, in time, be integrated with the study of population control into a comprehensive theory of mortality; and this, in turn, with population genetics.

Meanwhile, much neozoological work has been devoted to the dynamics of extant populations (see, in particular, Deevey (1947; Farner 1952; v. Haartman 1951; Hickey 1952). This is especially the case with avian populations, which, for several reasons, are more readily studied along these lines than many others. Data on the dynamics of other natural populations are as yet scarce, and this gap in our knowledge should be filled as soon as possible. Regarding mammals, for instance, most studies have concerned small forms, where the banding method is possible.

The program of "bigger and better life tables" holds great promise to the paleontologist, and perhaps in particular to the student of fossil invertebrates. In spite of this, paleontologists have not so far devoted much interest to the study of population dynamics, though much recent work— e.g., on growth stages—has a direct bearing on the subject. The lack of evolutionary—or, more broadly, biological—interest of many paleontologists was recently decried by Colbert and Newell (1948), and this state of matter is reflected in the partially more or less prescientific status of paleoecology today. Paleoecology has lost much of the stimulating contact with contemporary neoecology that Abel tried to give it, and at present, data excellently suited to modern ecological analysis are all too often left unheeded. Among the tools developed by neoecologists, that of the life table analysis may at least confer upon paleoecology a welcome addition of exact methodology.

THE LIFE TABLE

The construction of life tables may be carried out by three different methods (Merrell 1946, as summarized by Hickey, 1952). The dynamic life table ("horizontal life table" of Deevey 1947) summarizes the "survival and mortality of a given cohort . . . over a period of years." The time-specific, or "vertical" life table, is erected on census data from a single period of observation. The composite life table, finally, may rest on census or mortality data for different periods or different populations, the data being subjected to dynamic or time-specific analysis.

It is obvious that only the dynamic life table proper describes the actual dynamics of a cohort and that tables of the other types describe hypothetical populations. That this is so for the composite table is immediately obvious; that the same is true for the time-specific table becomes apparent if we assume that the population is increasing or decreasing in number, in which case the age structure of the census data will deviate from the temporal age structure of a cohort. Such differences assume great importance regarding populations that fluctuate widely in numbers, seasonally, or over a period of years. With less violently varying populations, however, this effect is of minor importance, and will tend to be reduced still more, e.g., if the data are of the census type and taken during several different periods: the result will then be a parameter, useful in description in about the same way as a mean for a variate. Long-range trends in population size will still make the time-specific table differ from the dynamic one, as is, for instance, the case with life tables for human populations; in the study of animal populations, however, the effect is likely to be overshadowed by small-sample effects, as the raw data are rarely very numerous.

The methods of gathering the raw data are, in principle, three: (1) to follow the fate of a cohort from birth to the death of the last survivor; (2) to observe the age at death for a random sample of the population; (3) a census of a living population or a random sample of it.

In paleontological work, the dynamic life table proper can seldom be used. A time-specific table may be constructed, if we find the remains of a population that was destroyed simultaneously. In most cases, however, only composite tables can be erected, and it is then a matter of judgment whether the treatment should take the form of a dynamic or a time-specific analysis.

This difficulty may be illustrated by analysis of a sample of the Silurian ostracod, *Beyrichia jonesi* Boll, from Gotland. Its ontogeny was described by Spjeldnaes (1951) from a fairly large sample, in which the successive instars could be separated. The initial two or three stages were under-represented, owing to the fragility of the juvenile tests. The ?fourth instar is the first adult one, and the frequencies from this stage on were:

Instar, Serial No.	Frequency
4	309
5	210
6	146
7	92
8	75
9	70
10	42
11	19

The data cannot be considered to pertain to a single period of observation, still less to a single cohort. Accordingly, a life table based on them is of the composite type. The analysis may be dynamic or time-specific; in the former case it is assumed that the remains indicate the age at death of the individuals; in the latter case the sample is considered to be a random sample of the population. Two different life tables may thus be constructed, as seen in table 1.21.

As life table treatment is new to paleontology (though the survivorship of taxonomic units through geologic time has been treated by the same method by Simpson 1944), a brief explanation may be inserted at this point; for a full discussion, however, the reader is referred to e.g. Pearl's

Table 1.21

Life Tables for the Silurian Ostracod, *Beyrichia jonesi* Boll, to Show Difference between Dynamic and Time-Specific Treatment. Raw Data from Spjeldnaes. Time in Terms of Moults, with Beginning at ? Fourth Instar. Sample Size 963 Specimens.

x age in terms of moulting stages	x' age as % deviation from mean longevity	d_x number dying in age interval	l_x number surviving at beginning of interval	$1000q_x$ mortality rate per 1000 alive at beginning of interval	e_x expectation of life or mean after life time (moults)
Dynamic					
4–5	−100	321	1000	321	2.38
5–6	−58	218	679	321	2.3
6–7	−16	152	461	330	2.1
7–8	26	95	309	308	1.9
8–9	68	78	214	365	1.5
9–10	110	73	136	537	1.1
10–11	152	43	63	682	0.8
11–12	194	20	20	1000	0.5
Time-specific					
4–5	−100	321	1000	321	2.62
5–6	−62	207	679	305	2.6
6–7	−24	174	472	369	2.5
7–8	15	55	298	185	2.7
8–9	53	16	243	66	2.2
9–10	91	91	227	401	1.4
10–11	129	75	136	552	1.0
11–12	167	61	61	1000	0.5

biometry text, and in particular to Deevey's most admirable exposition (1947).

The life table summarizes the fate of a cohort (real or imaginary), whose members start life together, by giving, for regular intervals of age, the number of deaths, the number of survivors remaining, the rate of mortality, and the expectation of further life, or the mean after lifetime. These columns are headed d_x, l_x, q_x and e_x respectively, x denoting age. The size of the initial cohort is given as 10^n, n depending on the size of the sample; in most ecological work the samples are between 100 and 1000 in size, and the cohorts are given as beginning with 100 or 1000 individuals. The rate of mortality is then usually put on a "per hundred" or "per thousand" basis, and headed $100q_x$ or $1000q_x$ respectively. The rate of survival is, of course, $1-q_x$.

A survivorship curve is obtained by plotting l_x (number of survivors remaining) against age x. This is preferably done on a semilogarithmic grid, l_x being on the logarithmic scale, and x on the arithmetic. In this way rates of mortality are reflected in the slope of the curve, regardless of the actual strength of the cohort.

The survivorship curves for the two interpretations of the *Beyrichia* data are given in fig. 1.15. As in the table, time *(x)* is measured in terms of moults, no absolute measure being available. In both cases the result is a curve of the "diagonal" type (Pearl and Miner 1935; see also below) with approximately constant rates of mortality irrespective of age; in a semilog graph this curve approaches a straight line. It is seen that the time-specific interpretation departs somewhat more from the ideal case than the dynamic, indicating decreased rates of mortality during middle life. But the mean rate of mortality per instar (M_w), calculated by means of the Magnusson and Svärdson (1948) formula,

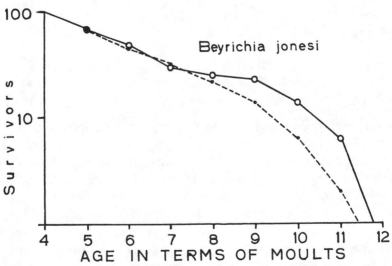

Fig. 1.15.   Survivorship curves for the Silurian ostracod. *Beyrichia jonesi,* based on dynamic (dots) and time-specific (circles) analysis respectively. Time (abscissa) in terms of moults.

$$M_w = \frac{D_1 + D_2 + D_3 + \ldots D_n}{D_1 + 2D_2 + 3D_3 + \ldots nD_n}$$

is identical (32.1 %) in both cases. Similarly, the first age group comprises 32.1% of the whole sample in both cases; thus the losses are exactly replaced, and both systems are in balance. As regarding the parameters, M_w and S_w (mean survival), dynamic and time-specific analysis will give closely similar results when the survivorship curve approaches the diagonal type, and identical in the ideal case (see also Hickey 1952, for a comparison between dynamic and time-specific treatment of data on birds). Mean longevity is slightly different in the tables, the dynamic analysis giving 2.38 moults, and the time-specific 2.62 moults (mean longevity being $= e_x$ at initial date).

The selection between the two interpretations is not immediately given, but Spjeldnaes (1951) thinks it most probable that the remains are largely of tests shed in moulting, and the sample would then be comparable with data of type (3) (census data); I am, accordingly, in favor of the time-specific interpretation.

DETERMINATION OF INDIVIDUAL AGE

In assembling the raw data, the determination of age is the crucial point. Here animals growing by moult stages, like the ostracods discussed above, may be profitably studied. Annual growth rings occur in the shells of some clams, and also in the genital plates of some sea-urchins (Moore 1935). Relative measures of age may be obtained from number of septa in chamber-shelled forms.

Regarding mammals the difficulties are often insuperable. To juvenile individuals the correct age may be assigned with some confidence from the development and cutting of the teeth, provided that the rate of development in the species is known or can be safely inferred. This method may work for adult specimens too, if the dentition is renewed in grown-up individuals, as in progressive proboscideans. Otherwise determination of individual age in adult animals is possible if there is some structure that is increased discontinuously, as the horny sheaths of some bovid horns (see e.g. Couturier 1938). Mostly, however, the horny sheaths do not fossilize. In general the morphological changes of a mammal are continuous during its adult life (wear of dentition, change of proportions), and, in such cases, individual age cannot be determined unless the specimens form a discrete age series (or other reliable standards are known). This will result e.g. if parturition is seasonal and the sampling limited to one season of the year. Also, the samples should be fairly large, if the distributional modes are to be recognizable.

Such conditions would seem to be very exceptional. Nevertheless there is abundant evidence that fossil samples meeting these requirements do occur, mammalian as well as others. Some of this evidence will be reviewed in the pages to follow.

Seasonal Deposition of Fossils

AGE GROUPS IN MERYCHIPPUS FROM THE SNAKE CREEK

Matthew (1924) has described a series of distinct ontogenetic age groups in the Miocene horse, *Merychippus primus* Osborn, from the Snake Creek fauna. The stages are as follows:

> 1. Milk premolars emerged but unworn; no permanent teeth. The stage is equal to that of a colt a few days or weeks old.
> 2. Milk premolars considerably worn; $M\frac{1}{1}$ emerged but almost unworn. Corresponding age in the horse somewhat more than one year.
> 3. Replacement of milk premolars proceeding DP: $^{2-3}/_{2-3}$ usually lost, whereas DP$\frac{4}{4}$ are present; permanent premolars not fully emerged; $M\frac{2}{2}$ emerged, $M\frac{1}{1}$ somewhat worn. Corresponding age in the horse between 2 and 3 years.
> 4. All permanent teeth emerged, $M\frac{3}{3}$ unworn. Corresponding age in the horse about 4 years.
> 5. All permanent teeth worn.
> 6. Teeth heavily worn, P_2 sometimes to root and lost.
> (Matthew 1924:163)

According to Matthew, the last two stages might be divided into four (probably more?), which merge into each other, but groups (1)–(4) are distinct, and there are no intermediates.

Matthew's interpretation is that these are annual growth stages in *Mer-*

ychippus, and he remarked on the fast ontogenetic evolution of this genus, as compared to that of *Equus,* "a three-year-old *Merychippus* being as fully grown as a four-year-old horse." This is obviously correlated with the small size of *Merychippus.*

Unfortunately no data on the variation within these stages are to be found in the literature, nor have I found any data on the number of individuals within each age group. The whole collection comprises "numerous skulls" (Matthew 1924), and the upper milk dentitions are about 200.

No other case of a similar grouping is reported from the Snake Creek, not even in *Merychippus paniensis* Cope, though this species is stated to comprise no less than 90 to 95 percent of all equids from these localities (Matthew 1924:159) and thus would seem to be represented by an enormous material. But *M. paniensis* belongs to another (younger) stratigraphic horizon than *M. primus.*

AGE GROUPS IN STENOMYLUS HITCHCOCKI FROM NEBRASKA

In a sample of the fossil camelid, *Stenomylus hitchcocki* Loomis, from the vicinity of Agate in Nebraska, Loomis (1910:303, 306) found an analogous series of growth stages.

(1) Complete deciduous dentition well worn, first molars emerging, but unworn. These young jaws were found "in surprisingly great numbers." The age of one year was assigned to these individuals.

(2) With permanent dentition, in which "the third molar is just up"; age, two years. Number not stated.

(3) Adults, number not stated.

Loomis thought it probable that the oldest individuals—with teeth worn to roots—were not more than five or six years old, as "every indication points to very rapid wear of the teeth" (first molars heavily worn between stages 1–2). This seems a reliable statement, considering that *Stenomylus hitchcocki* was a quite delicate creature (height at shoulders 684 mm. in the type). In this animal development was obviously much faster than in *Merychippus,* as there are only two juvenile age groups. But the age was probably somewhat overstated by Loomis. Stage (1) is about equivalent to stage (2) in *Merychippus* from Snake Creek, and stage (2) in *Stenomylus* to (4) in *Merychippus*. Thus the development of *Stenomylus* was just twice as fast as in *Merychippus,* and the age of group (1) in the former would be about six months, rather than one year. Moreover, if the age were one year, the animals would have died at about the time of parturition, and some newborn or foeta might reasonably be expected in this fairly wealthy sample ("not less than forty skeletons, and I should estimate many more"— ibid. p. 297). Consequently, the groups probably represent ½-year-olds, 1½-year-olds, and so on.

AGE GROUPS IN PERMIAN AMPHIBIANS

Seasonal deposition has also been indicated by Olson and Miller (1951) for samples of the Permian amphibians, *Diplocaulus magnicornis, Trimerorhachis insignis,* and *Lysorophus tricarinatus,* and possibly the Permian reptile, *Captorhinus aguti.* The frequency distributions of variates like skull length and central length are more or less clearly multimodal, showing distribution into several growth states—probably annual ones, though some influence of sex differences may be involved (ibid. p. 215).

In the *Diplocaulus* sample, the young of the year are represented together with at least two and probably more adult groups; the frequency is greatest within the first adult group. The *Trimerorhachis* sample most nearly approaches what may be the actual age structure of the population, for here the first group (young adults) is most numerous, the second markedly smaller, and the third very small.

AGE GROUPS IN THE HIPPARION FAUNA

Schlosser (1921) has discussed several cases of similar age grouping in the Old World *Hipparion* fauna. The Veles collection seems to be an especially good instance, and Schlosser, who compared the juvenile stages with similar ones in related recent mammals, went so far as to state that all the animals died in the autumn, probably in October.

Regarding *Hipparion* itself, Schlosser found growth stages in the Pikermi and Samos collections as well as in the Veles sample. The stages in the Veles material were as follows.

(1) DI_3 unworn, DP_{2-4} worn, M_1 emerging. Corresponding age in the horse somewhat more than one year. 9 individuals.

(2) P_{2-4} well worn, P_4 emerging. Somewhat more than 3 years in the horse. 1 individual.

(3) Not distinct older stages, at least 12 individuals. In two cases comparison with recent horses indicated an age of 13 and 14–15 years respectively.

The *Hipparion* colts from Pikermi are partly foetal or newborn, partly again 10–12 months old, in which M_1 has emerged. The number of individuals is not stated.

The *Hipparion* material of Samos permits a more extensive tabulation of the age groups. According to Schlosser, they would be about as follows (though his treatment is very involved and sometimes obscure).

(1) Foetal or newborn with unworn milk teeth, all of which are emerged. This group forms "einen bedeutenden Bruchteil" of all the fossils.

(2) $M\frac{1}{1}$ emerging, "an mehreren Kiefern": 10–12 months in the horse.

(3) M^2 emerging. 20 months in the horse. 1? individual.

(4) M^2 functioning, $M\frac{3}{3}$ emerging, $P^{2-3}/_{2-3}$ replace the milk premolars, whereas $DP\frac{4}{4}$ are still retained though heavily worn. $2\frac{1}{2}$ years in the horse. Possibly 3–4 individuals.

(5) $I\frac{2}{2}$ emerging, DI^3 and DC still functioning, P^2 almost unworn, P_{2-4} emerged, M^3 emerging, $3\frac{1}{2}$ years in the horse, 2? individuals.

(6) Adults, number not stated.

The comparisons with the horse are Schlosser's, and it can be seen that the age difference between the successive stages averages less than one year, or about 10 months. Schlosser discussed the possibility that the age grouping was a result of two (or more) catastrophes; the first three stages might represent animals killed by one catastrophe, and the remaining two, then, would be the victims of another. This hypothesis is of course at variance with Occam's Razor, and Schlosser concluded, undoubtedly correctly, that the rate of development of *Hipparion* was different from that of *Equus:* "Ob sie (conditions in *Equus*) ohne weiteres auf *Hipparion* übertragen werden dürfen, erscheint insoferne zweifelhaft, als *Hipparion,* weil von

geringeren Körperdimensionen, wahrscheinlich frühreifer war als *Equus*
. . ." There is here a queer error, probably a slip of the pen, caused by
confusion of data; Schlosser was more concerned with the season of the
catastrophe than with rates of development. If—as implicitly stated by
Schlosser—the stages in *Hipparion* were separated by one year of development, and the corresponding stages in *Equus* by ten months, the conclusion would be that the development of *Hipparion* was *slower* than that of
Equus.

This is however very unlikely, and, as will be shown below, the Chinese
evidence demands a different interpretation. My opinion is that groups (4)
and (5) are in fact identical, both comparable to horses of the age of about
3½ years, and that the ages, in terms of horse development, of the pre-
ceeding groups are stated too low. This would imply that a *Hipparion* of
three years of age was as fully developed as a 3½-year-old horse. Similar
conditions were found in the Chinese *Hipparion*.

AGE GROUPS FROM THE PONTIAN OF CHINA. HIPPARION

The *Hipparion* sample of the Lagrelius collection is probably not a
homogeneous one, though the extreme splitting into 11 species carried out
by Sefve (1927) is unlikely to reflect the actual number of specifically
differentiated populations represented. The collection is sampled from
several localities, scattered over a wide area in Northern China. Yet, surprisingly enough, the juvenile age groups are clearly recognizable. The
fossils at different localities are certainly not contemporaneous in origin,
but they were clearly deposited at the same season of the year, even if the
temporal gaps between the deposition of different fossil pockets may
amount to thousands of years.

The juvenile stages are largely similar to those described by Schlosser
(1921) from Veles, Pikermi, and Samos, but I find it impossible to carry
out any distinction between his stages (4) and (5). The grouping of the
Chinese material is given below; data on the development of *Equus* are
from Martin's (1914) tables.

(1) Deciduous premolars emerged but completely or almost unworn, no permanent teeth. 0–2 months in the horse. 18 individuals.

(2) Deciduous premolars worn, first molar emerged but unworn.
About one year in the horse. 13 individuals.

(3) Deciduous premolars worn to roots, replacement proceeding, second molar emerging, rarely worn. 2½ years in the horse. 8 individuals.

(4) Third molar emerging, completely or almost unworn. At least 3½ years in the horse. 7 individuals.

The conclusion is that the Chinese *Hipparion* developed at a somewhat higher rate than the recent horse; a three-year-old *Hipparion* was almost as fully developed as a *Merychippus* of the same age. The same is probably true for the *Hipparion* populations from Samos, Pikermi, and Veles, the rate of development thus being correlated with the size of the animal.

CHILOTHERIUM

As with *Hipparion,* the *Chilotherium* sample of the Lagrelius collection is probably not homogeneous, though it is likely that taxonomic splitting (Ringström 1924) has been brought too far. Like the *Hipparion* fossils, those of *Chilotherium* are derived from several localities, though geographically from a relatively small area, most being from the neighborhood of Pao-Te-Hsien. Again the juvenile stages are clear-cut and easily recognized, as follows.

(1) Foeta with emerging unworn $DP^{2-3}/_{2-3}$; no other teeth were found. 4 individuals.

(2) $DP^{2-3}/_{2-3}$ in function but almost unworn; DP4 beginning to emerge, but mainly concealed in the jaw. Very young animals, probably not fully weaned. 9 or 10 individuals.

(3) $DP^{2-3}/_{2-3}$ considerably worn, DP4/4 emerged but little worn or not worn at all. M1/1 concealed in the jaws. Upwards of 18 individuals.

(4) $DP^{2-3}/_{2-3}$ heavily worn, M1/1 emerging, rarely worn. About 13 individuals.

(5) All milk teeth worn to roots, replacement beginning; P_{2-3} often well emerged. M2/2 emerging. 10 individuals.

(6) P2/2–M2/2 in function, M3/3 emerging. About 13 individuals; this age group is possibly heterogeneous.

(7) Older animals, age groups obscure. 25 individuals or more.

Parturition occurred between stages (1) and (2), and probably not long after (1), as the young of stage (2) seem to have been weaned, in part at least. Probably parturition occurred about a quarter of a year after stage (1).

Here a study of growth may be inserted to see whether this grouping is consistent with regular growth processes. A number of variates could be followed through the main part of the development: the width across the palate at P^3; the width of the lower jaw at the symphysis, anterior to P_2, and at P_4; and the height of the ramus, before and behind the first and last teeth of the tooth row, respectively.

The sample being heterogeneous taxonomically, there is considerable variation within each stage; but there is a steady increase in most dimensions, retardation in some being outweighed by acceleration in others, indicating that the grouping is valid (fig. 1.16). When the growth curves are reduced to a common scale (in percentage of maximum), some differences in the growth patterns emerge. Thus the symphyseal width is positively allometric, being also the only dimension not wholly retarded in growth during the sixth year.

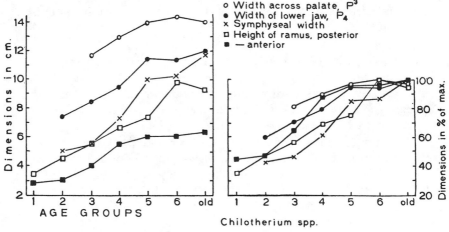

Fig. 1.16. Growth of various dimensions in *Chilotherium* spp., as labelled; means within successive age groups. At least two or three species combined. Left, absolute data; right, dimensions expressed as a percentage of maximum size, permitting comparison between rates of growth.

SAMOTHERIUM

The *Samotherium* sample, from several localities, is certainly heterogeneous; Bohlin (1926) describes at least 4 species. But again the juvenile specimens fall into distinct age groups.

(1) Young animals, probably less than 1 year old: $DP^{2-4}/_{2-4}$ in function, but little worn or not worn at all. 5 individuals.

(2) DP considerably worn, first molars emerging. 13 individuals, most of which (7) are from Loc. 30.

(3) DP heavily worn, second molars emerging. 6 individuals.

(4) Replacement finished, third molars unworn or slightly worn. 3 individuals.

(5) Two or three individuals showing a stage of wear somewhat advanced beyond that of group (4).

(6) Older specimens, 17–18.

For a study of growth only one variate was considered suitable, the maximum width of the ramus, as the remains are often very fragmentary. The heterogeneity of the sample would seem to induce marked irregularities in the ensuing growth pattern. Yet the growth curve (fig. 1.17) is remarkably smooth, with the exception of a dip in Group 5, which is represented by two specimens of the rather small form, *S. neumayri*.

PLESIADDAX DEPERETI[7]

Among the larger samples in the Lagrelius collection, that of the ovibovine, *Plesiaddax depereti* Schlosser, is unique insofar as it is almost completely derived from one fossil pocket, Loc. 114 North, near the town Ho-Ch'ü in Shansi. The subsequent analysis of the age groups in this sample is based on upper dentitions, which are more numerous than lower ones. Four specimens are from Loc. 114 South, one from Loc. 305 (in the vicinity of Pao-Te, Shansi), and the rest from 114 North.

The young individuals are easily separable into distinct age groups without intermediates (see fig. 1.18).

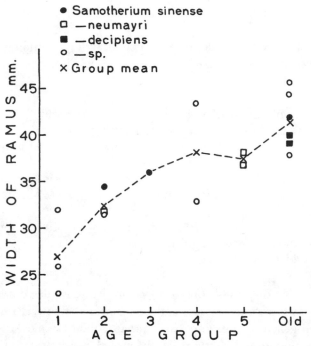

Fig. 1.17.   Growth of ramus width (maximum) in *Samotherium* spp; single observations and means per age group, as labelled. Twin observations on single individual recorded as a mean.

(1) Milk premolars unworn or slightly worn, first molar likewise. Second molar not emerged, but formed in the jaw. No trace of permanent premolars.

(2) Milk premolars heavily worn, second molar emerged but unworn or nearly so, third molar formed in the jaw. Permanent premolars partly or fully formed in the jaw (see Bohlin 1935; Pl. VII, fig. 10).

(3) Tooth replacement finished, permanent premolars and third molar slightly worn.

From the presence of the clear-cut juvenile stages it is fairly safe to infer that the adults are similarly divided into annual age groups. The identification and separation of these is however a delicate task, involving careful statistical analysis.

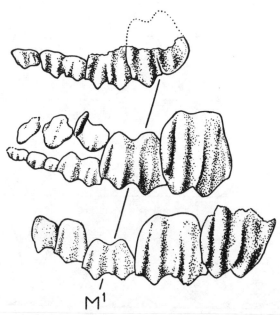

M^1

Fig. 1.18. Juvenile stages in development of upper dentition of *Plesiaddax depereti,* one year apart. Top, with unworn M^1; center, with permanent premolars forming in jaw; bottom, with permanent dentition (P^2 missing). Note heavy wear of first molar. Redrawn after Bohlin. Two-thirds natural size.

That is not a hopeless task, however, is apparent from the way in which the presence of age groups was first detected; this was not done by inference from the juvenile age groups, but the other way around. This will be briefly described, as it seems to strengthen the point. When studying the relative wear of the cheek teeth from an extensive table of crown heights, I was occasioned to graph the values, as in fig. 1.19. It was then noticed that the data clustered around certain values, with gaps in between. The conclusion was obvious, and it was vindicated by a study of the juvenile specimens.

The existence of age groups being granted, the task is to find out the grouping that is most natural. The method employed was to measure the height of every cheek tooth, molars and premolars, of *P. depereti* in the collection. Distribution diagrams (fig. 1.19) indicated the number of groups present and the means of crown height for each group. It was helpful to find that the fourth stage, with somewhat worn M^3 and premolars, could be singled out on purely biological criteria (aspect of denti-

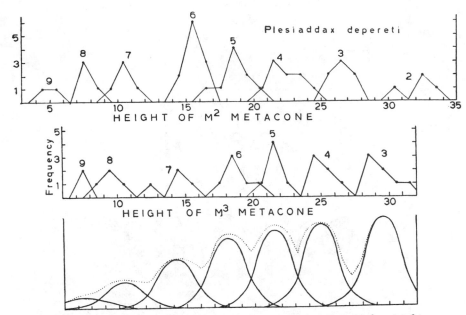

Fig. 1.19. Top and center: frequencies of metacone heights of M^2 and M^3 in age groups of *Plesiaddax depereti*. Bottom: Diagrammatic representation of frequency distributions of M^3 height in successive age groups, showing overlap of variation between groups. The multimodal combined curve shows obliteration of groups in oldest individuals.

tion) and thus gave additional information on the general amount of wear per annum.

Fig. 1.19c suggests how the method works in principle and illustrates its potentialities and limitations. It shows, diagrammatically, distribution curves for the height of M^3. The range of variation of each age group overlaps that of the preceding and succeeding ones; the combined curve is multimodal, the modes being progressively obliterated in the oldest specimens.

This leads to a seeming self-contradiction, to wit: whereas the frequency within each age group may be estimated, no specimen after stage (4) (which is recognizable on biological criteria) can be assigned to its actual age group with absolute certitude. Thus, for instance, a specimen at the mean of Group (6) may in reality be a borderline case of Groups (5) or (7). Such borderline cases are however equally liable to occur in both of two adjacent groups, and to about the same extent, as the groups are generally subequal in size, and the variation within successive groups is of a similar order of magnitude. As a fact, these two factors tend to balance in the issue: in any two successive groups, the older one tends to have the lesser frequency and the higher variation. Hence, for instance, the number of Group (5) members incorrectly assigned to Group (6) will be theoretically balanced by about an equal amount of Group (6) members incorrectly assigned to Group (5). The relatively small number of specimens in the present case of course tends to increase the chance fluctuations, but this does not invalidate the result.

In the graphs, the frequencies appear to be highest in the middle groups, (5)–(6). This does not reflect the actual frequencies. The lower number of observations in Groups (2)–(4) is due to the fact that the molars in these groups are not fully emerged and their bases cannot be reached without preparation. Some specimens were prepared in the course of Bohlin's work, but as these age groups could be determined on other criteria, and preparation involves damage to the material, I deemed it unnecessary to continue preparations. The same is true, *mutatis mutandis,* for the younger age groups in other Bovidae.

The results were checked by renewed reference to the specimens and careful weighting of the indication given by the apparent state of wear of the teeth. The grouping was determined from the whole dentition, but the indication of the second and third molars (which have higher crowns and wear more rapidly than the others) was held to be the most important. It was found that the great majority of specimens fitted well into the groups.

There are a few borderline cases though. In one instance the right and left M^3 gave different indication. This was probably the result of delayed emergence on one side, and the indication of M^2 was followed. More difficult to place were some specimens in which the hind part of the dentition is lost or damaged; the estimate from the premolars is uncertain, and at least two groups might be probable. The specimen was then assigned to that group (out of the two or three possible ones) which was most evidently under-represented. This final resort was taken to in four cases only and seems justifiable.

The grouping is consistent with the fact that tooth wear was very rapid in these animals (see fig. 1.18, showing excessive wear of M^1 in the two-year-interval between first and third stage). It should be stressed however that the three oldest age groups are less objectively delimited than the others. It can be seen from fig. 1.19 that variation overshadows the group distinctions, and the number of specimens is small. There may be more than three groups, but in my opinion this is the most probable number, considering the rate of wear of the dentition in earlier stages. The frequencies are similarly uncertain, but it is evident that the number of individuals shrinks rapidly in these old groups, and my interpretation seems to express this fairly.

The juvenile development of *P. depereti* proceeded about similarly to that of domestic sheep, perhaps at a slightly slower rate, and the conclusion is that the first stage (equivalent to sheep of 5 months) represents lambs about 6 months old. The ages and frequencies in the *Plesiaddax depereti* sample are then, according to my estimate, as follows:

AGE IN YEARS	FREQUENCY
½	11
1½	5
2½	13
3½	12
4½	12
5½	10
6½	7
7½	4
8½	2

The total is 75, an increase of 11 individuals from my estimate in 1952. The more refined methods of the present census have caused me to refer to different age groups left and right dentitions which otherwise might have belonged to the same animal. This does not, of course, affect the conclusions in my previous paper, which were based on relative minimum frequencies. The same applies to the other samples, as expected.

URMIATHERIUM INTERMEDIUM

The ovibovine *Urmiatherium intermedium* Schlosser was about the same size as its ally, *Plesiaddax depereti,* or very slightly larger; it is however decidedly more hypsodont. The remains are from several localities, but the age groups are as clear-cut as in *Plesiaddax,* and the juvenile stages in both are almost identical. A certain lag in development may however be detected in Groups (2) and (3); the third molar of *U. inter-medium,* in Group (3), is generally completely unworn, whereas in *P. depereti* the paracone is touched by wear. The conclusion is that development in *U. intermedium* was slightly retarded in comparison with *P. depereti,* but the first stage probably represents lambs of 6 months in this case too.

The adult groups could be separated in the same way as in *P. depereti* (see fig. 1.20), and the same remarks apply to them. In this case, as in the former one, the upper dentition was studied; the frequency of calvaria or maxillas is markedly greater than that of lower jaws in both species.

According to my estimate the *U. intermedium* age groups number one more than those of *P. depereti,* or 10 groups, with frequences as follows:

AGE IN YEARS	FREQUENCY
½	19
1½	10
2½	16
3½	15
4½	12
5½	11
6½	10
7½	10
8½	5
9½	2

The dentition was worn down in about the same manner in both species. Fig. 1.21 shows annual means for crown heights. Table 1.22 summarizes the wear of the molars. The wear of each tooth is most rapid at the beginning, when the free cones are abraded. A difference between the species is seen in the fact that the fourth premolar seems to have been more worn

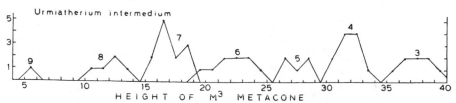

Fig. 1.20. Frequencies of M^3 metacone heights in age groups of *Urmiatherium intermedium.*

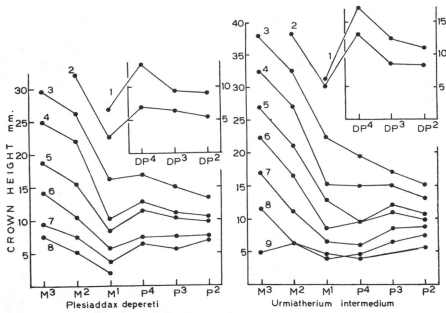

Fig. 1.21. Annual means for crown heights in upper dentitions of *Plesiaddax depereti* (left) and *Urmiatherium intermedium* (right).

during life in *U. intermedium* than in *P. depereti;* possibly a functional (though not structural) molarization of this tooth had begun in the former species.

The annual wear of the horse molar has been estimated at 2.2 mm. (Ellenberger and Baum, 1926) or 2.5 mm. (Lundholm 1947). Presumably these data pertain to horses that pasture on less harsh food than that available to these Pliocene ovibovines; also, it should be remembered that the grinding surface of the equine dentition is larger in relation to body size.

GAZELLA DORCADOIDES

The gazelles of the Lagrelius collection *Hipparion* fauna probably comprise at least three species, of which one *(G. dorcadoides)* is certainly and another *(G. gaudryi)* probably polytypic (see Bohlin 1939, Kurtén 1952). One important character is degree of hypsodonty. The most hypsodont species is *G. dorcadoides* Schlosser, but this species is evidently a compound

Table 1.22
Age Groups in *Gazella dorcadoides*

Age in Years	Frequencies			
	Loc. 30	Loc. 43	Loc. 109	Total
¾	9	2	4	15
1¾	13	6	11	30
2¾	11	2	7	20
3¾	5	3	5	13
4¾	3	3	4	10
5¾	1	2	2	5
6¾	0	0	1	1

of several subspecies, most of which are confined to one or a few localities; at each locality, however, there is generally not more than one subspecies.

A study of the age groups of *G. dorcadoides,* hence, cannot be carried out on the species as a whole, since the samples differ in hypsodonty. Instead some rich localities were selected and studied independently. I have selected Locs. 30, 43, and 109. The *Gazellas* of the two former quarries are of the same type and seem to stand very near the type of *G. dorcadoides* Schlosser. Those of Loc. 109 are somewhat larger and could perhaps be referred to the subspecies (species?) that Bohlin (1935) tried to identify with *G. altidens* Schlosser. Whether this is a temporal or geographical subspecies I do not attempt to determine.

I tried to make a similar study of the gazelles from Loc. 114 too, but this turned out not to be feasible, for there are here two forms, one probably near or conspecific with *G. paotehensis* Teilhard and Young, and the other probably a relatively brachyodont *G. dorcadoides* subsp., slightly larger and somewhat more hypsodont than the former.

Age groups:

(1) Milk dentition well worn, first molars in function and worn, second molars emerging or just beginning to wear. At Loc. 114 this stage is more advanced, for here the third molars are also fully formed and sometimes beginning to emerge. This would indicate either that the "catastrophe" of this locality occurred somewhat later in the season, or else that the time of parturition of these gazelles was somewhat earlier.

(2) Full adult dentition, third molars not much worn.

(3)–(7) Adult dentitions in various stages of wear. These stages were separated in accordance with crown height of upper and lower molars, as seen from fig. 1.22.

Comparison with recent bovids of the same size would seem to fix the time of parturition some eight or nine months before stage (1), the ages being roughly ¾ year, 1¾ year, and so on (see table 1.23)

The oldest individual on record would thus have attained an age of about seven years. It is a striking fact that the number of individuals drops at a rapid and uniform rate from the second group on, in contrast with the age groups of *Plesiaddax* and *Urmiatherium;* the significance of this will be discussed later on.

Table 1.23
Annual Abrasion of Molars (in mm.) of *Plesiaddax depereti* and *Urmiatherium intermedium*

	Plesiaddax depereti		
Age Group Interval	M^1	M^2	M^3
1–2	4.1		
2–3	6.8	6.5	
3–4	5.6	3.8	4.3
4–5	2.6	6.8	6.1
5–6	3.1	5.2	4.7
6–7	1.7	2.8	4.4
7–8	2.1	2.5	2.1

	Urmiatherium intermedium		
Age Group Interval	M^1	M^2	M^3
1–2	1.2		
2–3	7.4	5.8	
3–4	7.3	5.6	6.7
4–5	2.2	6.1	5.4
5–6	4.4	4.5	4.7
6–7	2.4	3.1	5.4
7–8	1.6	5.3	4.9
8–9	0.7	–	6.9

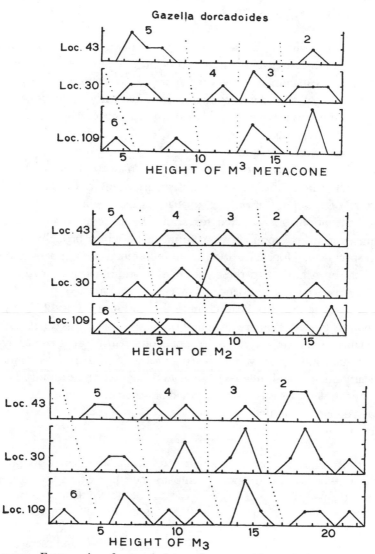

Fig. 1.22. Frequencies of crown heights of upper and lower molars in age groups of *Gazella dorcadoides* from localities 30, 43, and 109. Note slight superiority in hypsodonty of sample from Loc. 109.

ICTITHERIUM WONGII

Something is already known about the population structures of herbivorous mammals (see below), but very little about the vital statistics of carnivorous mammals. It was therefore of great interest to find that the ictitheres of the *Hipparion* fauna were similarly divided into age groups as the herbivorous mammals discussed above. The groups were found to be very distinct and clear-cut, even though the remains are derived from several different quarries.

The *I. hyaenoides* sample is somewhat too small to permit of accurate study along these lines, though investigation indicates that the age structure of this population was probably closely similar to that of *I. wongii,* the population selected for study. Some hitherto underscribed specimens, touched upon above were included, which brought the total to 40 specimens. Upper dentitions are predominant in numbers, and the study has been limited to these, except that additional information was sought in associated lower dentitions whenever possible.

The stages appear very beautifully in the amount of wear of the upper carnassial (fig. 1.23). Primary grouping of the specimens was carried out in accordance with the indication of this tooth. The precarnassial teeth were studied subsequently, and the stages of wear in these were found to be clearly distinct and fully consistent with the present grouping. Hence it was possible to classify some fragmentary specimens, which lacked the carnassial, fairly accurately.

The stages are as follows.

(1) Recently mature individuals, probably some 8 or 9 months old, in which the permanent premolars have just emerged and are almost unworn; faint traces of wear can be detected on the metacone and metastyle rims of the P^4. Canines not fully emerged. 10 specimens.

(2) All teeth slightly worn, face on metastyle blade and metacone rim about 1.5 mm. broad. The face on P^3 was measured in two directions (anteroposterior and transverse diameters) and the product, roughly proportional to the area of the face, was about 6.0 mm^2. 9 specimens.

(3) Wearing faces broadened, that of P^4 metastyle to 3–4 mm; P^3

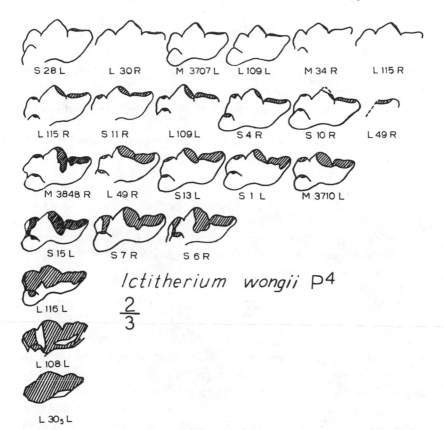

Fig. 1.23. Upper carnassials of *Ictitherium wongii*, lingual view, grouped according to wear (wearing face hatched). All two-thirds natural size. S, specimen No., M, museum No., L, locality No.; L (at the end), left; R, right (reversed).

product 19.0. Faces on metacone and metastyle of P⁴ still apart. 7 specimens.

(4) Teeth markedly worn, faces on metacone and metastyle of P⁴ flowing together, wearing face forming groove from paracone to protocone. P³ product 26.1. 5 specimens.

(5) Continuous face of wear on lingual side of P⁴, connecting all cusps; wearing area on paracone much enlarged. Profile of tooth distorted from wear. P³ product 32.5. 3 specimens.

(6) Only small parts of P⁴ retain enamel. P³ product 38.9. 2 specimens.

(7) P^4 completely abraded on lingual side, whole dentition badly worn. 2 specimens.

It should be stressed that groups (5)–(7) are less distinct than the younger ones, and the grouping is tentative.

The stages are seemingly consistent with the following notations:

(1) The wear is rather slow as long as the resistant enamel is not pierced; after that the wearing faces are rapidly enlarged;

(2) These small hyaenids probably did not attain more than seven or eight years of age under natural conditions, since the large recent hyaenas in captivity, at about 20 years, show signs of old age and have teeth completely worn out (Flower 1931), though probably enjoying more easily masticated food than under natural conditions;

(3) The frequencies within the groups are reduced at about a similar rate throughout the series.

CAUSES OF SEASONAL DEPOSITIONS OF FOSSILS

As to the causes of this "seasonal sampling"—which has been exemplified from Snake Creek, the *Stenomylus* beds, Veles, Samos, Pikermi, and the Chinese *Hipparion* fauna—most authors favor the flood theory (Loomis 1910, Schlosser 1921, Kurtén 1952). Matthew (1924, 165), however, wrote somewhat contemptuously about "the traditional textbook idea that fossil remains were animals drowned by rivers during flood, whose remains were washed down by the stream to their present position." His opinion on *Merychippus primus* was that the animals died during a few weeks of drought, and their remains were accumulated around the few remaining pools of water—a theory that from an actuo-geological point of view is very well founded.

This theory would, however, force us to the conclusion—highly unsatisfactory from an ecological point of view—that *Merychippus primus* foaled at the worst time imaginable. The youngest individuals in the collection are almost newborn: they were then born shortly before "the few weeks of low water" . . . "when the stream was reduced to a string of scattered pools to which the animals resorted to drink and were destroyed by carnivora, caught in quicksands or met with other misadventures" (ibid.).

During the first weeks of their life, then, the colts would have been at the mercy of the most disadvantageous conditions of the whole year, and the females would have had to produce milk during the driest season. This would indeed be an adaptive valley.

In zones where the climate differs strongly from one season to another, parturition generally takes place shortly before conditions become most favorable. This is probably a result of acute adaptation to environmental conditions and governed by strong selection pressure. Even in a well adapted species, mortality is very high during the first period of infancy, and lack of adaptation in this respect would probably lead to extinction.

In spite of these objections it seems to be an inescapable conclusion that seasonal catastrophes of some kind did occur at Snake Creek, Veles, Samos, Pikermi, and so on, at the time of parturition of *Merychippus* and *Hipparion* respectively. Schlosser even states that the initial age group of *Hipparion* may be partly comprised of foeta.

A possible, and indeed probable explanation would be that these catastrophes were outside the normal order of things—that they were either local or else they did not occur annually.

Regarding Veles, Schlosser points to a local catastrophe as a possible cause of death: "Wie ich ... erfuhr, ist die Gegend von Veles auch heutzutage nicht allzu selten der Schauplatz von verheerenden Wolkenbrüchen und in deren Gefolge von reissenden Wildbächen und Überschwemmungen." A more or less local catastrophe is also suggested by Abel (1912) in his classical Pikermi theory: the animals tumbled down a precipice when fleeing before a steppe fire. The fractured limb bones are good evidence; the steppe fire though should probably be modified into a wood fire (Kurtén 1952: 31). Of course a preceding period of drought must be taken into account.

As to the Snake Creek accumulations, the theory of Matthew cannot be refuted without very good reason. With his eminent knowledge of the area in question this great paleontologist was an authority whose opinions demand careful consideration. Indeed the sampling may be due to a belated and abnormally severe period of drought—once, or several times: i.e., a combination of unfavorable conditions that may have been rare but of a devastating effect. A recurrent, "normal" period of drought some time before parturition is not at variance with the ecological considerations. A population is probably far less affected by environmental rigors before, than after, parturition, as can be judged from the fact that, in most Northern mammals, gestation coincides with the most unfavorable climatic season of the year.

Regarding the Chinese *Hipparion* fauna it seems appropriate to consider the various possibilities in turn and see to what extent they fit the facts. The facts are, briefly resuméd:

(1) Seasonal deposition has been ascertained for most or all of the dorcadoides and mixed fauna localities, which are widely distributed over three provinces in Northern China. These localities have most species in common, and no fact decisively counters the view that deposition occurred during a period in which there were no major changes in the composition of the fauna.[8] (2) The season of the year in which deposition predominantly occurred seems to have been the same throughout. (3) There are however striking differences in the relative frequencies of the species. A species which at one locality is predominant in numbers may be represented at other localities by one or a few specimens. Almost every rich quarry is characterized by at least one such mass occurrence.

Factors possibly involved in seasonal mass deaths may be biotic or abiotic. To the former ones belong predation and diseases; to the latter ones, drought, cold, fire, floods and the like.

The biotic factors need not detain us for long. In the case of predation, why does it occur at the same season at widely different places and times? Why the simultaneous death of various forms, probably with different enemies, and the presence of the predators themselves? Some epidemic, on the other hand, might conceivably kill off a great percentage of a local population in a short time, and it might even recur at a fixed season; but the facts then demand a peculiar sort of epidemic, at one time killing *Plesiaddax* and a few unfortunate members of other populations, and at another time concentrating on gazelles, chilotheres, and ictitheres.

Apart from this, however, the biotic factors are ruled out because of the simple fact of mass fossilization as a sequel to mass death. It would indeed seem that such simultaneous mass deaths of mammals are fairly rare phenomena in nature. But similarly fossilization is a rare phenomenon. Both require quite extraordinary environmental conditions. The coincidence is really unlikely to occur *unless* the "extraordinary circumstances" destroying large numbers of animals are, in some way, *conductive also to fossilization*. Wiman (1913), indeed, held the radical—probably too radical—opinion that "auf dem Lande dürfte das gewöhnliche Sterben für die palaeontologischen Funde gar keine Rolle spielen." Abel (1912) gives several instances of mass deaths that may lead to fossilization; *inter alia,* the interpretation of the volcanic ash deposits of the basal Bridger formation as resulting from a single eruption—Pompeii on a grand scale.

The evidence in the present case—cumulative from the large number of observations—strongly suggests that the factors causing mass deaths and mass depositions were identical or strongly correlated; and the almost inevitable conclusion is that the streams, which are responsible for deposition, were somehow connected with the deaths too. This rules out, e.g., cold as the decisive factor.

Drought is then more plausible. It may serve to collect a variety of animals to the remaining pools, and those that die there may later be deposited into the sediments of the river, when the flow again increases. But in that case the remains in the quarries apparently should be random samples of all animal populations inhabiting the neighborhood. Drought is a menace to every population, not to just species A in one area and to species B in another. In samples as large as many of the present ones the relative frequencies have great statistical significance, and the conclusion would be that the territories in the vicinity of e.g. Loc. 114 North—so to speak, the uplands from which came the fossils of Pool 114 North—were almost exclusively populated by *Plesiaddax depereti;* whereas plesiaddaces at other places tolerated the drought better. The samples are clearly *biassed* in composition.

"The traditional textbook idea" appears to provide a better explanation; and it is not quite so much out of touch with realities as Matthew insisted. In his *Principles of Sedimentation* Twenhofel (1950) states: "Conditions on floodplains . . . are not always conductive to preservation, as rates of deposition may be slow and burial may take a long time," but "the occasional flood may kill large numbers of animals in a short time and bury the carcasses in the deposits of the channels."

The reconstruction, then, would picture middling to large, generally intraspecific herds of animals—perhaps with a few members of other species mixing in, as is common, e.g., with the African wild life of the present day: perhaps a herd of urmiatheres with some occasional antelopes associating; a pack of ictitheres, and so on. The mass occurrences are always of species the modern counterparts of which are gregarious, whereas, e.g., the sabre-tooths seem to have been solitary. The flood comes; and what happens depends on sheer luck: one herd is overtaken by the water, another reaches safety in time. Under such conditions, the drifting carcasses, at last coming to rest in some backwater where most of them are collected, will be the remains of a few associations of animals within the sweep of the flood. And as such they appear in the quarries: *not* as random samples drawn from every population in the neighborhood.

Loomis, who favored the same explanation regarding the *Stenomylus* beds, pointed to the peculiar death position, seen in many instances, e.g., in the type of *S. hitchcocki*. The head is thrown back, as if in a desperate attempt to catch a breath of air; "the position . . . I believe is common among drowned animals" (Loomis 1910: fig. 1.). In one case only is an articulated portion of the skeleton represented together with the skull in the Lagrelius collection of the *Hipparion* fauna. This is a young *Chleuastochoerus stehlini*, figured by Pearson (1928, Fig. 25), seemingly in its death position, which is similar. The atlas of some carnivores has been found in juxtaposition with the occipital condyles; the impression received is that of a cramped position, but this evidence is slender. However, this position is very common in dying animals and does not necessarily indicate drowning.

Yet the .flood theory seems to fit the facts better than the previously discussed ones; but, e.g., a steppe fire, driving a herd of animals into the stream, may have had similar results. Both possibilities were tentatively suggested by me (Kurtén 1952), and I do not think the problem definitely solved, though analogy with the present-day floods of the Hwang-Ho furnishes valuable accessory indication.

The remains may probably be considered as random samples (with some qualifications, to be discussed later on) of herds, and thus of the populations, insofar as the herds were actually random samples. This cannot *a priori* be stated to be so, for there might have been herds made up of adolescent males, or by a patriarchal male with a suite of females and their young, and so on. With few exceptions, however, there is no indication of such differential associations.

This protracted and, perhaps, somewhat meticulous discussion is a necessary prelude to the study of the population dynamics in these mammals. It has shown, convincingly I believe, that the correct method in this instance is the time-specific analysis, the raw data being of the census type.

It may finally be pointed out that, though many mass occurrences result from seasonal deposition, not all do, not even when there is some kind of selection favoring the the preservation of certain species. The immense numbers of dire wolves and sabre-tooths mired at Rancho La Brea do not seem to form discrete age series, as far as can be judged from the literature (Merriam 1908, 1909; Merriam and Stock 1925, 1932). A somewhat similar case was reported in the newspapers in the summer of 1952. In the course of draining works at a parsonage in Norway, it was found that the bottom of the unfortunate minister's well was covered with numerous rat carcasses, resting on top of a sediment of rat skeletons. This might have been a promising future fossil quarry.

Comparative Life Table Treatment

STATE OF PRESERVATION AND ITS INFLUENCE ON APPARENT AGE STRUCTURE

The fossils of the Chinese *Hipparion* fauna are mainly remains of the skeleton of the head: whole skulls, sometimes with jaws *in situ*, parts of skulls and jaws, often isolated maxillas or broken rami. Remains of post-cranial skeleton are less common. Limb bones have been found in some cases (notably a wealth of samothere limb bones), vertebrae are rare, and there is no complete articulated skeleton in the collection. The previously discussed, partial *Chleuastochoerus* skeleton is an exceptional case.

This, as well as the nature of the deposits (red *Hipparion* clay) indicates conditions of sedimentation somewhat differing from those of the *Stenomylus* beds (evenly sorted and bedded sands, according to Loomis 1910), in which complete skeletons, evidently in the position of death, are common.

The disarticulation of the remains is probably to a great extent due to the action of running water. The paucity of remains of small mammals indicates that the streams were rapid, and an amount of smaller fragments were probably washed away, whereas the *Stenomylus* sandstones, "it would appear . . . were laid down in some more sheltered area behind a barrier" (Loomis 1910). The result is that the juvenile age groups, in which the sutures of the skull were not yet closed, are markedly under-represented.

This may also to some extent have been caused by the action of scavengers; some bones are evidently gnawed by hyaenas.

In the larger forms, with slow development, these features appear particularly clearly. The first age group in *Chilotherium* (foeta) is represented by isolated rami only. In stage (2) there appear isolated tooth rows, but also complete mandibles; evidently the mandibular symphysis had reached a considerable degree of strength. In stage (3) there are already some complete skulls, and this is the age group of maximal frequency; adult size is not yet attained, but the skull bones were closely enough connected to keep the calvarium together in many instances. The frequencies may give relatively good indication of the age structure of the population from this stage on.

The juvenile underrepresentation recurs through the whole species list. In *Gazella*, stage (1) is under-represented; in stage (2) the animals are fully grown. In *Plesiaddax* and *Urmiatherium* the two first stages are juvenile, and underrepresented; in stage (1) there are no complete calvaria, in stage (2) a few. The main part of the remains consists of isolated dentitions. In the third and following stages there is a high percentage of more or less complete calvaria. A striking feature is that most good skulls are male. In both species there is strong sexual dimorphism; in this respect *Plesiaddax* is probably not surpassed by any living or extinct mammal (see Bohlin 1935). The females of this species were hornless, the males had short and thick horn-cores, the bases of which merged into mighty domes of bone. Probably the presence of these structures endowed the skulls with great strength and did much to keep them from becoming disarticulated. The female skulls, lacking these characters, were probably broken into pieces to a much greater extent, and in the cases where the fragments include portions that permit sexing, they generally turn out to be female. It is thus probable that the majority of the fragmentary specimens are females and that the apparent preponderance of male skulls is spurious. As to the juveniles, it may be noted that the one nearly perfect young *Urmiatherium* skull is that of a 1½-year-old male, with massive horn-cores forming.

The frequencies of juveniles are thus not directly comparable to those of adults. It seems therefore appropriate to discuss adult and juvenile population dynamics separately.

LIFE TABLES FOR FOSSIL AND RECENT MAMMALS

The most complete life table for a natural mammal population available today is that computed by Deevey (1947) on Murie's data, for the Dall mountain sheep, *Ovis d. dalli*. The material amounts to more than 600 skulls, which were collected in the field, and the predominant cause of death was assumed to be predation by wolves. The age at death was determined from the annual rings on the horns. The analysis was, of course, dynamic.

As observed by Deevey, however, and emphasized in Bourlière's (1951) criticism, the juvenile classes (lambs and yearlings) are underrepresented in the collection, and hence, in Deevey's table, juvenile mortality appears lower than it actually is. The order of magnitude of this bias will be discussed later on.

This underrepresentation is doubtless due to the perishability of the tender skulls, as with the fossil samples; the bones are less massive, and the sutures being open, the remains are easily disarticulated, scattered or destroyed by predators and scavengers. From the age class 1–2 years on, however, when maturity is attained, the data seem to be reliable, i.e., "the probability of finding a skull is not likely to be affected by the age of its owner" (Deevey 1947). Consequently, the adult part of the life table should be reliable. As the same remarks apply to the fossil samples, the *Ovis dalli* life table should provide an important standard for comparison, and I have accordingly recast this table, with beginning at age class 1–2 years (table 1.24).

Bourlière (1951) has computed a table giving the frequencies within the annual age groups in the chamois (*Rupicapra rupicapra*) sample on which Couturier's (1938) monograph is based. The sample is evidently in the main a collection of hunting trophies. Hence it includes very few specimens of less than 2 years of age. The adults, on the other hand, may be assumed to be killed at about the same rate irrespective of age. To be sure, hunters are keen on large specimens with magnificent horns, which spells old males, and this might tend to increase the relative frequencies of the older age groups to some extent, while having the opposite influence on the living population. It must be remembered that the chamois is a game animal under close human persecution, which may tend to distort the results; on

Table 1.24
Life Tables for Two Recent Mammal Populations, Based on Data from
Murie and Couturier. Adults Only; Both Sexes Combined.

Species: Size of Sample	x	x′	d_x	l_x	$1000q_x$	e_x
Ovis d. dalli	1–2	−100	15	1000	15.0	7.7
(497)	2–3	−87	16	985	16.5	6.8
	3–4	−74	15	969	15.5	5.9
	4–5	−61	37	954	39.3	5.0
	5–6	−48	58	917	62.6	4.2
	6–7	−35	60	859	69.9	3.4
	7–8	−22	86	799	108	2.6
	8–9	−9	165	713	231	1.9
	9–10	4	233	548	426	1.3
	10–11	17	195	315	619	0.9
	11–12	30	113	120	937	0.6
	12–13	43	3	7	500	1.2
	13–14	56	4	4	1000	0.7
Rupicapra rupicapra	2–3	−100	(184)	1000	(184)	3.67
	3–4	−73	(156)	(816)	(191)	3.4
(422)	4–5	−45	139	658	211	3.1
	5–6	−18	212	519	409	2.8
	6–7	9	47	297	158	3.5
	7–8	36	74	250	296	3.0
	8–9	63	74	176	421	3.1
	9–10	91	9	102	88	4.0
	10–11	118	0	93	0	3.3
	11–12	145	37	93	398	2.3
	12–13	172	19	56	339	2.5
	13–14	200	18	37	486	2.5
	14–15	227	1	19	53	3.5
	15–16	254	9	18	0	2.5
	16–17	282	(8)	(18)	(445)	1.5
	17–18	309	1	10	100	1.5
	18–19	336	9	9	1000	0.5

the other hand its natural enemies are much reduced in its natural habitat. These factors may, or may not, cancel out. Regarding birds there are indications "that samples obtained by shooting are biased with respect to young birds, i.e., shooting takes a greater proportion of the young than of older birds" (Farner 1952). There are thus several sources of possible bias. They would not seem, however, to invalidate the main features of the life table (Table 1.24), which is computed by time-specific analysis. This analysis gives $M_w = 24.0\%$ (the Magnusson and Svärdson formula). Dynamic analysis of the data gives a closely similar result, or $M_w = 25.7\%$.

These life tables are given as for cohorts of 1000 individuals, the size of the samples exceeding 100 specimens. The tables (table 1.25) for the fossil populations are given as for cohorts of 100 individuals, the samples never exceeding this number. Small-sample effects are visible, but the general

Table 1.25
Life Tables for Early Pliocene Mammal Populations. Adults Only; Both Sexes Combined. Original Data.

Species; Size of Sample	x	x'	d_x	l_x	$1000q_x$	e_x
Urmiatherium intermedium (81)	2.5	−100	6	100	6	4.6
	3.5	−78	19	94	20	3.8
	4.5	−57	6	75	8	3.7
	5.5	−35	6	69	9	3.0
	6.5	−13	1	63	2	2.2
	7.5	9	31	62	50	1.2
	8.5	30	18	31	58	0.9
	9.5	52	13	13	100	0.5
Plesiaddax depereti (60)	2.5	−100	8	100	8	4.1
	3.5	−76	0	92	0	3.4
	4.5	−51	15	92	16	2.4
	5.5	−27	23	77	30	1.8
	6.5	−2	23	54	43	1.4
	7.5	22	16	31	52	1.0
	8.5	46	15	15	100	0.5
Gazella dorcadoides (79)	1.75	−100	33	100	33	2.1
	2.75	−52	24	67	36	2.0
	3.75	−5	10	43	23	1.7
	4.75	43	16	33	49	1.1
	5.75	90	14	17	82	0.7
	6.75	138	3	3	100	0.5
Ictitherium wongii (38)	0.75	−100	10	100	10	3.3
	1.75	−70	20	90	22	2.6
	2.75	−39	20	70	29	2.2
	3.75	−9	20	50	40	1.9
	4.75	21	10	30	33	1.8
	5.75	52	0	20	0	1.5
	6.75	82	20	20	100	0.5

features are probably valid. The selected initial date always represents the first adult stage.

SURVIVORSHIP CURVES

The survivorship curves for these fossil mammals, and for the two recent species, *Ovis dalli* and *Rupicapra rupicapra*, are shown in fig. 1.24, together with a typical survivorship curve for a recent avian population (the blackbird, *Turdus m. merula*, after Deevey, data by Lack). As in fig. 1.15, the plotting is semilogarithmic, with age (in years from the selected initial date) on an arithmetic scale, and the number of survivors on a logarithmic scale. This makes possible a direct comparison between rates of mortality, which are reflected in the slope of the curve. The impression of the actual shrinkage between successive age groups is of course distorted; fig. 1.25, showing age pyramids for the same populations, shows this aspect without distortion.

Pearl and Miner (1935) discussed several theoretically possible types of survivorship curves, the main ones being the negatively skew rectangular (convex), the diagonal (straight) and the positively skew rectangular (concave). In fig. 1.24, a good diagonal is represented by the blackbird,

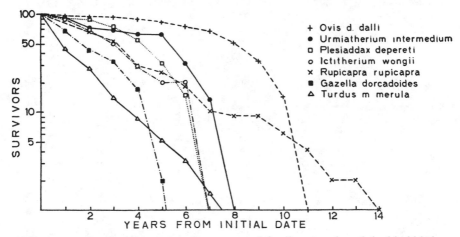

Fig. 1.24. Survivorship curves for recent and fossil mammals and the blackbird, *Turdus m. merula*, as labelled. Adults only. Ages at initial date: *Ovis dalli*, 1 year; *U. intermedium* and *P. depereti*, 2½ years; *I. wongii* ¾ year; *R. rupicapra*, 2 years; *G. dorcadoides*, 1¾ year.

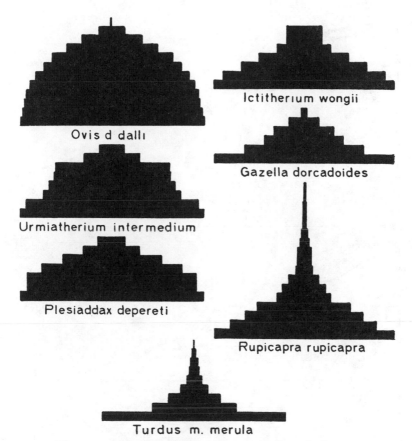

Fig. 1.25. Age pyramids for the same species as in fig. 1.24.

and a fair approximation to the convex type by *Ovis dalli.* Of the remainder, the chamois approaches the diagonal type, and the two ovibovines the convex type, whereas *I. wongii* and *G. dorcadoides* are intermediate.

From fig. 1.24, a comparison between absolute rates of mortality in the various populations may be carried out, but a direct comparison of the *forms* of the survivorship curves necessitates a different scaling. Age should then be expressed in terms of percentage deviation from the mean duration of life (in this case, from expectation of life at initial date), a device developed by Pearl and Miner.

Fig. 1.26 shows the result. Two curves recur as standards in fig. 1.26–

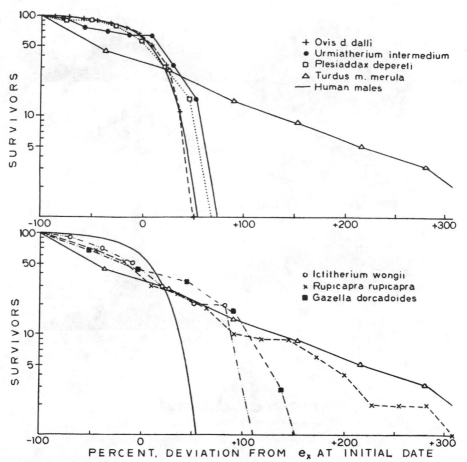

Fig. 1.26. Survivorship curves as in fig. 1.24, but with age expressed in percentage deviation from mean longevity. Top, "convex-type" curves; bottom, "diagonal-type" and intermediate curves.

that of the blackbird as an instance of the diagonal type, and that of the human male (continental whites from the U.S.A., beginning with subadult age class 14–15 years; data from Allee et al. 1949) as representing the convex type. All are comparable insofar as they record the adult life spans of the various organisms in question.

The top half of fig. 1.26 records survivorship curves of the convex type, and these are seen to be very closely coincident in spite of great differences in absolute longevity, ecology, geological age, and taxonomic position. The Dall sheep curve is almost exactly confluent with the human curve, and the deviations in the curves for *P. depereti* and *U. intermedium* might conceivably result from small-sample effects. In all four cases, the last survivor out of a cohort of 100 adults would attain an age of 1.5–1.75 times the mean duration of life.

The more or less diagonal survivorship curves in the bottom half of fig. 1.26 present a greatly different picture. In both the blackbird and the chamois, the life span of the last survivors vastly exceeds the mean longevity of the adults. The slope, in the ideal diagonal-type curve, corresponds to a mortality of about 32% per 100% deviation (Deevey, 1947); with an initial cohort of 100 adults, the line intersects the age axis at about +340%. Both populations are close to this pattern.

The ictithere and gazelle populations take an intermediate position. In both the main part of the curve follows a course closely similar to the first half of the curves for the blackbird and the chamois; but at about +100% deviation there sets in a rapid elimination of the survivors, and out of a cohort of 100 adults, the last survivor would seemingly die before +150% deviation. The data serving as a basis are not very numerous, and it might be argued that this convexity to the tail of the curve is really an artefact: the actual *Ictitherium* cohort begins with 10 specimens, and in a sample of that size the last survivor should die at +120% deviation, if the curve is of the diagonal type; in a cohort of 100 adults the life span would be greatly extended. The numerical data do not thus suffice to demonstrate a real deviation from the diagonal curve type. But when the state of utter decrepitude of the dentition in the oldest specimens is considered, it is quite clear that survivors at +100% deviation stood a very small chance indeed of extended survival. The downturn of the curve is certainly valid. The same holds, *mutatis mutandis,* for the other fossil populations as well, where the numerical data *per se* are sufficient to demonstrate the convexity.

The curves for the annual rates of mortality in the same populations are shown in fig. 1.27. The curves for human males and for the blackbird again serve as standards for comparison. Aside from irregularities, which are related to the size of the sample, all convex curves (1.27 top) show a similar distribution of mortality rates, low at the beginning of adult life and mounting steeply towards the end of the life span. The diagonal-

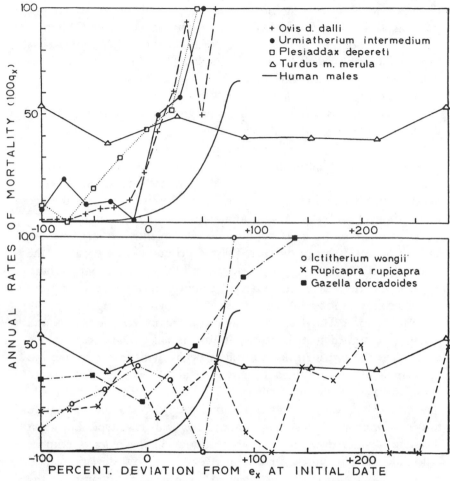

Fig. 1.27.   Rates of mortality; for explanations see fig. 1.26.

type curves (blackbird and chamois, 1.27 bottom) are characterized by nearly constant rates of mortality, those of the chamois indeed fluctuating widely, but with a marked central tendency. In the ictitheres and gazelles, the rates are more or less constant initially, later to increase steeply.

Expectation of life is represented in fig. 1.28 and the difference between the convex-type and the diagonal-type curves, as well as the intermediate position of the ictitheres and gazelles, is again apparent. In the convex-

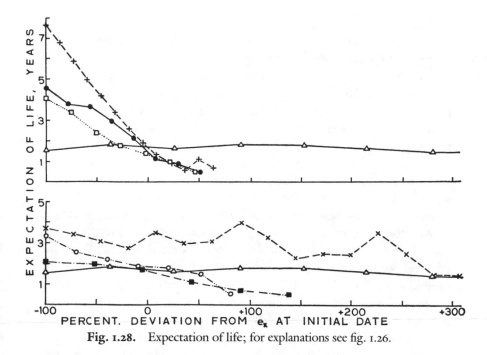

Fig. 1.28.   Expectation of life; for explanations see fig. 1.26.

type population, life expectancy is at a peak initially, and is reduced at a fairly uniform rate; in the diagonal-type populations, e_x is constant throughout. The blackbird is an almost ideal case, but the chamois curve, though irregular, fluctuates with a central tendency, around a mean slightly above 3 years. In the gazelles and ictitheres, finally, there is a very slow reduction of life expectancy with increasing age.

JUVENILE MORTALITY

The collection of Dall sheep skulls, in Deevey's analysis, indicates a juvenile mortality of about 20 percent. On the other hand, the usual productivity of sheep is one young per annum and adult female, with occasional twin births. Older females are probably barren; to what extent this may balance the twin births is not known, but for the sake of simplicity a mean productivity of one young will be assumed. The production of lambs in one year would then be about thrice the number in the initial age group of Deevey's life table. It may thus be inferred that juvenile mortality is radically higher than 20 percent, the present estimate indicating a value of about 70 percent. This is almost certainly nearer to the actual amount, though it is of course quite possible that the potential productivity of the population is not realized in full.

The productivity of the chamois is one young per annum and breeding female; according to Couturier (1938), twin births are exceedingly rare. This indicates a juvenile mortality of 55 percent, or considerably less than in the Dall sheep. The difference is hard to account for. The fact that man has killed off a majority of the natural enemies of the species, and is himself the main enemy of the adult chamois, may have brought about an increase of adult mortality rates and a decrease of juvenile rates; such intercompensation has been demonstrated in other cases (e.g., Errington 1946). But there is another possibility that cannot be staved off, viz., that the productivity does not compensate for the loss, and that the chamois population is actually shrinking in numbers, or was so when the sample was collected.

On the other hand, even higher rates for juvenile mortality in mammals than the estimate for *Ovis dalli* are known. Thus Green and Evans (1940) calculate the loss, between birth and the following February in the snowshoe hare (*Lepus americanus*), as 77 percent. This does not however differ much from the adult annual rate, which is given as 70 percent.

The fossil samples give some rough indication of the order of magnitude of juvenile mortality. Thus, in both *Plesiaddax* and *Urmiatherium*, a shrinkage of about 50 percent is indicated between ages 6 and 18 months; in *Hipparion*, about 46 percent during the second year of life; in *Chilotherium*, 28 percent between ages 21 and 33 months, and 23 percent between 33 and 45 months. These results are to some extent corroborated by an estimate of the total juvenile loss, when the probable productivity of the species is taken into account.

Thus the age structure of the *Hipparion* sample is:

AGE IN YEARS FREQUENCY

AGE IN YEARS	FREQUENCY
0	18
1	13
2	7
3	8
4 and over	108

Assuming a productivity of 1 colt per annum and adult female, the initial group should number 54, instead of 18. The underrepresentation would seem to be progressively reduced in the older groups, and the cohort may probably be taken to have entered the age of 4 years with some ten survivors, which gives a juvenile mortality of no less than 81 percent.

In the *Urmiatherium* sample there are 10 1½-year-olds and 16 2½-year-olds. If the former quantity is tentatively corrected to 17 individuals, the strength of the ½-year-group would be about 30 individuals. It must probably be assumed that rates of mortality were still higher during the first half year of life. These data point to a productivity higher than 1 young per annum and female, and an average of 1½ seems to be the minimum necessary for a stable population. Under such circumstances the total amount of juvenile mortality, from newborn to 2½ seems to be the minimum necessary for a stable population. Under such circumstances the total amount of juvenile mortality, from newborn to 2½-year-olds, would be on the order of 74 percent.

An analogous estimate for *Plesiaddax* seems to necessitate the assumption of still higher productivity, on the order of 1¾ per female and annum, giving a juvenile mortality rate of about 75 percent.

In *Gazella dorcadoides,* the productivity may have been comparable to that of the recent *G. dorcas,* where twin births are normal (Brehm-Ekman 1938); this would give a juvenile mortality of about 60 percent. The productivity of *Ictitherium wongii* was probably about as in larger recent viverrids, or 2–3 young; juvenile mortality may have been 70–80 percent.

The *Chilotherium* collection is probably biassed (see Ringström 1924, Kurtén 1952), and the analysis cannot be carried further.

The age structure of the *Samotherium* sample is:

AGE IN YEARS	FREQUENCY
0	5
1	13
2	6
3	3
4	2
5 and over	18

This would indicate a steadily reduced rate of mortality, from about 54 percent in the second year through 50 percent in the third to 33 percent in the fourth. The distribution of the older dentitions into age groups is not possible to determine with any exactitude, owing to the heterogeneity of the sample. *Samotherium* was a large and presumably long-lived animal, and so the annual groups would seem to average three or four specimens at most. A more or less subjective estimate can of course be formed, and it would then seem that the adult samotheres were rather evenly distributed over the older age groups, up to an age of perhaps 15 years or more. The data are much too uncertain to warrant publication, but the indication is that the age distribution of adult samotheres was strongly convex. A stage with thoroughly worn teeth is attained by at least four individuals, which may be members of the two or three oldest age groups.

In this case, comparison between the frequencies of adults and juveniles in the sample seems to indicate that the juveniles are in excess, especially the 1-year group. The main part of this is derived from Loc. 30 and may represent a non-random association.

For juvenile mortality in various populations, the following more or less uncertain values may thus be listed:

Hipparion spp	81%
Lepus americanus	77%
Ictitherium wongii	75%
Plesiaddax depereti	75%
Urmiatherium intermedium	74%
Ovis dalli	70%
Gazella dorcadoides	60%
Rupicapra rupicapra	55%

Though the consistency is fairly good, it would certainly be rash to conclude that the average of 7/10 is valid, as the calculations leading to this datum are riddled with subjective assumptions. Many more data are urgently needed, in particular from field observations of living populations.

As they stand, the estimates have been used as a basis for the synthetic survivorship curves of fig. 1.29, where birth is taken as the initial date.

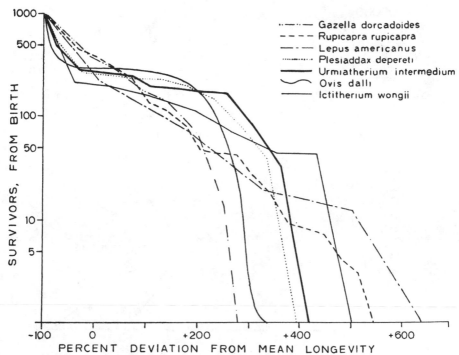

Fig. 1.29. Synthetic survivorship curves (from birth) for recent and fossil mammals.

In view of the uncertainties which hamper almost every step in the analysis, it need hardly be emphasized that these curves are not proposed to describe faithfully actual conditions. The intention is only to arrive at a rough estimate of the general form of the curve. Full life table analysis is certainly not warranted.

In all curves there is a pronounced dip initially; it appears to be most emphatic in curves of the convex type, but is surprisingly slight in the curves for the chamois and *Gazella*. This is also true for the snowshoe hare (this curve is taken from Deevey 1947, the data being from Green and Evans).

These curves may be compared with that for the prawn, *Leander squilla* L. (fig. 1.30). This is also a synthetic curve, based on a set of data for the Gullmar Fiord population published by Höglund (1943). The age ratios of the adults were based on censuses of catch in 1940 (males) and 1941

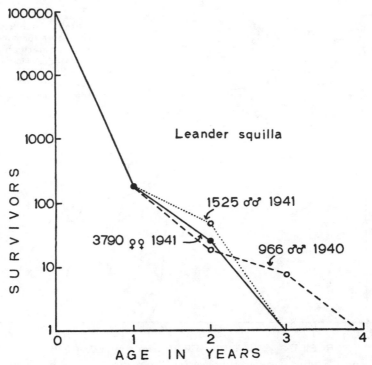

Fig. 1.30. Synthetic survivorship curves (from egg) for the prawn, *Leander squilla,* based on catch as labelled; data from Höglund.

(males and females). Very small and very large prawns are less easily caught than middle-sized ones, and the set of data is thus not fully satisfactory. The ratios between the one-year-olds and two-year-olds seem, however, fairly reliable. The productivity may be ascertained from Höglund's data with good accuracy.

The curve shows a tremendous dip in the first year, with a mortality rate of 99.8 percent, and seems later to be stabilized with a diagonal pattern, though data for prawns older than 2 years are uncertain. Adult mortality rates average somewhat short of 90 percent.

THE GENERAL FORM OF THE SURVIVORSHIP CURVE

We may now return to Pearl's and Miner's theoretical concept of three basic patterns of survivorship: convex, diagonal, and concave. The synthetic curves do not in any instance conform fully to any single one of these patterns, though indeed the *Leander* curve approaches the concave type, and would probably do so still closer if full data on the distribution of mortality were available for the first year of life.

The rate of mortality seems always to be high in the first period of life. This is true even for the human survivorship curve (see, e.g., Allee et al. 1949), and this would seem to be a feature that cannot be obliterated even if medical science were advanced far beyond its present state; for in fertilization there will almost inevitably appear some lethal or semilethal combinations which are weeded out by selection. Thus the initial dip is a constitutional character, in sexually reproducing forms at least, and it cannot be annihilated unless adult mortality rates, for other reasons, are higher than juvenile rates. This is probably uncommon, since in most cases juvenile individuals are more vulnerable than adults. It should however be pointed out that, where metamorphosis occurs, the animal is shifted into a wholly new individual-environment relation, and there is no *a priori* reason for supposing that this will always result in decreased rates of mortality (consider, in particular, cases like the 17-year cicada, or the Mayfly). Again, as Deevey (1947) observes, high rates of mortality may continue in later life as a neotenic character.

The initial concavity of the curve may at any rate be considered as a recurrent feature of fairly general validity. During adult life, on the other hand, conditions are variable. Except for the "neoteny" mentioned above, a common trend seems to be a gradual increase of mortality rates, characteristic of the convex-type curves (man, Dall sheep, and so on). This would seem to be due to what Pearl terms "endogenous senescence" in conjunction with increased vulnerability. But this trait is probably not universal. There are several data suggesting—though not definitely proving—that survival is increased with age in some populations with predominantly diagonal-type survivorship, notably avian (Farner 1952). The factor at work may be experience (as suggested by Farner), increase of hardiness and resilience as a result of selection (as suggested by Macdonell, 1913, to account for the high expectation of life in old age groups of ancient Romans), or, if the trend be proved to exist in invertebrates that grow throughout life, the survival value of larger size.

During the main part of adult life, then, the survivorship curve in broad outline may be straight, slightly convex, or slightly concave. This is probably not always correct in detail, for mortality may not always be equally distributed within each interval. Seasonal variations may be conspicuous, and it is highly probable that e.g. the adult *Leander* curve may actually be a series of steps rather than a smooth diagonal.

A rigid distinction between convex and diagonal survivorship patterns cannot be upheld. The potential life span is restricted for every organism, and, provided that the population is large enough, a fraction will inevitably reach an age where senescence eliminates the last survivors. There will thus always be a convexity to the tail of the survivorship curve, though it may concern an almost negligible fraction of the cohort (Farner, 1952). This is indeed suggested by the intermediate type of the *Ictitherium* and *Gazella* curves, which to some extent bridge the gap.

The basic pattern of the survivorship curve would therefore seem to be that represented by Pearl and Miner in their subtype B_2, a sigmoid curve, concave initially, fairly straight during middle life, and convex at the end. This will of course be subject to various distortions which may all but annihilate some of its properties.

I do not propose the abandoning of the classification of Pearl and Miner, which is perfectly useful. It is however clear that different segments of a curve may belong under different headings, and that there are no hard and fast boundaries.

The generalization may possible be brought one step further. Where very extensive data concerning convex-type survivorship patterns are in hand, the curve can be shown to terminate in a final concavity, with reduced rates of mortality (in man, in various laboratory populations, e.g. *Proales,* and starved *Drosophila,* as recorded by Pearl and Miner, and seemingly in the Dall sheep). This may be explained by senescence having a probability distribution, which carries a fraction of the cohort beyond mean potential longevity; under ideal conditions, as set up theoretically by Bodenheimer (1938), this distribution would be given by the first derivative curve of the survivorship curve.[9]

Aspects of Selection

SELECTION AND THE SURVIVORSHIP CURVE

Selection may be broadly defined as differential reproduction (Simpson 1949). As such, it may be, and commonly is, adaptive, at least in some way; but some kinds of selection may clearly be definitely detrimental biologically, as with some male epigamic characters enhanced by intra-sexual selection (see Huxley 1942:484). Another instance may be discussed here, as it is related to a subject that will later be more thoroughly discussed.

The implications, in terms of selection, of the commonplace fact that large females of a fish species tend to carry more spawn than small ones, are not always realized. Yet this is evidently a case of selection for larger size. The instance is not isolated. Rensch (1947) has shown that large poikilothermous vertebrates tend to have larger broods than smaller-sized allies, and points to the significance of this for the phenomenon of phyletic growth. That the same factor is also at work *within* populations has been demonstrated by Petersen (1950), who found that the relation between size of mother (cephalothorax length) and number of offspring in lycosid spiders can be described by the formula,

$$y = bx^k,$$

the values of k varying between 2.53 and 2.99.

A similar relation exists in crustaceans; I have graduated it for some species (see fig. 1.31). The data on the Gammaridae are from Spooner

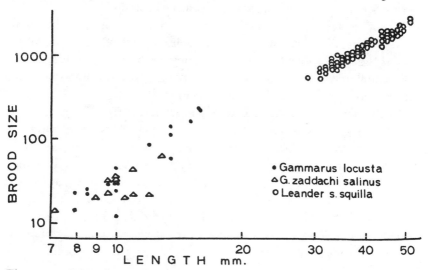

Fig. 1.31. Dependence of brood size on body length in Crustacea, as labelled: data from Spooner and Höglund.

(1947) and those on *Leander* from Höglund (1943). For *Gammarus locusta* L., $k = 3.53$, for *Leander s. squilla* L., $k = 2.76$ in young individʳⁿᶦs, and 3.43 in older ones. The data on *G. zaddachi salinus* Spooner are less numerous, but they indicate a relation essentially similar to that in *G. locusta*. It appears that this relation is identical or closely similar in the gammarids and the prawns.

In all these cases, the factor would seem to bring about a selection pressure towards increase of size, and the fact that the forms seem to be stable is seemingly due to this tendency being offset by some kind of counterselection (see Petersen 1950 for some suggestions). It may be tentatively suggested that the present gross size is optimal for the current mode of life of the species, but it can probably be concluded that, if the counterselective factor were put out of action, a phase of rapid phyletic growth might ensue. The trend toward size increase is very common in invertebrates, as in vertebrates (see Newell 1949), and in part at least it may be caused by the factor under consideration. It is interesting to note this as an explanation, in strictly selectionist terms, of a phenomenon that has long been advocated as evidence for an "inner urge." It seems that the vitalist idea has a core of truth insofar as the pressure is nonadaptive or preadaptive rather than strictly adaptive. It would seem to offer a possible mechanism of "released spring action" for some instances of quantum evolution.

If not at fertilization, the adverse selection countering this trend should express itself as differential mortaility during the life span of the cohort. Of the various factors contributing to the complex of selection, differential mortality certainly ranks among the most important and among those profoundly affecting adaptation.

As the survivorship curve changes in character from interval to interval in the life history of a cohort, so, presumably, the nature of selection is altered. What may be a primary aspect of this change relates to the intensity and effectiveness of selection. From the mathematical studies of population geneticists (see, for instance, the review by Wright 1949) we know that the effectiveness of selection depends on the size of the population on which it acts. In this light, the simple fact of successive shrinkage of a cohort with increasing age takes on a great significance. Selection pressures of equal strength may differ greatly in effect when directed towards the numerous initial age groups of a cohort, and towards its dwindling, old age groups. When such pressures differ in direction, even to the extent of opposing each other—conditions that may prevail when larval stages are greatly different from adult ones, but also where no radical metamorphosis occurs—the direction favoured in the young will gain, even if it is to the detriment of the very old.

In "prodigal" species, in the terminology of Deevey—those with enormous output of young—the tremendous mortality in early stages seems, however, to be largely nonselective. This does not invalidate the selection principle for the remaining fraction (Huxley 1942), but it results in lower selection pressures for the age class as a whole.

To some extent the selection in the young and that in the adults is probably different, even where metamorphosis does not occur. Thus abiotic environmental factors may be more important in early stages than in the less vulnerable adults (Bodenheimer 1938); the young may have specific predators, and so on. It might be useful to distinguish juvenile-specific and adult-specific mortality factors, recognizing, however, that the difference cannot be fully specified in many cases, and that many factors are identical in both stages. Thus, as regarding the lethal genes, probably responsible for part of the initial dip of the survivorship curve, it should be added that the lethal factor is really an extreme variant within a graded series of adaptational values, and thus the difference is one of degree.

With the final downturn of the survivorship curve, there enter into the picture the selective factors of senescence, "exogenous" and "endogenous" which may however be responsible for part of the mortality in earlier stages as well. At the end of the life span, these will generally affect the evolution of the species only if the age groups still partake in breeding, though in social forms nonbreeders may be of importance in selection. The effect of selection in these age groups will be greatly reduced where only a small fraction of the population attains old age, partly because it constitutes a limited percentage of the parents, and partly because of the relaxation of selection in small populations. The evolution of *Gryphaea*, as interpreted by Simpson (1944), is a case in point. The progressive coiling is regarded as an adaptation to life in mud, and death by self-immurement came only to extreme variants in their old age. In forms with convex survivorship patterns, however, factors increasing span of life and length of reproductive period will clearly have selective advantage.

Many of these aspects of selection defy analysis so far. But a few of them may clearly be evaluated to some extent on the basis of data already at hand.

HYPSODONTY AND POTENTIAL LONGEVITY

"Potential longevity is the maximum life span possible for an individual of a species under optimal conditions. Potential natural longevity is the maximum life span possible for an individual of this species under natural conditions." (Farner 1952). In the following, potential longevity will be denoted P.L., and potential natural longevity P.N.L.

The difference between P.L. and P.N.L. is clearly brought forth if the Pliocene Bovidae considered here are compared with, for instance, the chamois. The latter may live to an age of 22 years (Bourlière 1951), whereas the life span of e.g. the urmiatheres—which were considerably larger than the chamois—probably did not exceed 10 years. A chamois of 10 years is still physically fit and able to reproduce; and there is no reason for doubting that the same was true for the urmiathere. The exception concerns one character only: the teeth are worn to the roots, and the animals would soon have died from inability to masticate their food. P.N.L. of these Pliocene bovids was apparently limited by the wear of their dentition, and Flower (1931) finds this to be true for many recent mammals as well.

It has been mentioned already that tooth wear was very rapid in these herbivores of the dorcadoides fauna. The first molars of the Bovidae, for instance, are generally worn down to less than half of their initial height during the relatively short period up to the emergence and functional inset of the third molars (see figure 1.18) The same was evidently true for *Stenomylus hitchcocki,* as Loomis (1910) states: "the wear is very rapid, for when molar 3 is but slightly worn the first molar is down to the bottom of the pit."

A remarkable contrast is seen in the chamois. "L'usure des molaires et des prémolaires est encore beaucoup plus lente" (Couturier 1938). From the excellent X-ray photographs (pls. 24–25) and comments in Couturier's monograph it appears that the first molar is but barely visibly abraded

during the same period of development. Couturier's comments are pertinent to the present discussion: "Mais il faut insister sur un phénomène
important: l'usure très lente des dents; car, chez le Chamois, celles-ci
continuent à pousser une grande partie de son existence. Les autres Rum-
inants, le Chevreuil, le Cerf, le Daim, le Mouton, usent leurs dents beaucoup plus vite. Or, la dentition joue un rôle indéniable dans la longévité,
ce qui laisserait supposer que notre *Rupicapra* pourrait parvenir à un âge
assez avancé."

This difference presumably reflects differences in feeding habits, and the
indication is that the vegetables available to the inhabitants of the
Pontian steppe were of a tough brand indeed. Unfortunately no plant
remains have been found, but Ringström's (1924) observations on *Chilotherium* may be quoted. "Trotz der bedeutenden Hypsodontie sind
die Zähne aller älteren Exemplare bis auf die Wurzeln abgekaut, was zeigt,
dass die Nahrung Härterer Art war . . . An einer Anzahl Hauer . . . ist an
der Innenseite eine Unzahl feiner, paralleler Riefen zu sehen, die recht
wohl durch Schliff durch ein kieselsäurehaltiges Grass entstanden sein
können." The basic food of the chamois, on the other hand, is the Alpine
trefoil (*Trifolium alpinum*) and some other succulent plants. Consequently the strain on the dentition is slight, and P.N.L. is greatly
increased.

P.N.L. should obviously change with ecological conditions. A chamois
that had to resort to tough steppe grass like that available to the
Bovidae of the dorcadoides fauna, would probably wear out its dentition
in less than 10 years, and an urmiathere living exclusively on *Trifolium*
might have a chance to outlast the oldest chamois on record.

When ecological conditions impose heavy strain on the dentition,
differences in hypsodonty may be of importance in survival. An analysis
of the conditions in the Bovidae of the dorcadoides fauna gives some
information in this respect.

Urmiatherium intermedium was slightly larger than *Plesiaddax depereti*:
for instance, crown length of M^3 is about 30 mm. in the former and about
29 mm. in the latter. There is however a marked difference in hypsodonty,
the indices (100 crown height/crown length) being 127 and 102 respectively.
From the data on annual wear of teeth (see above, p. 64), means for the
annual wear of M^3 in each species may be computed. This mean was found
to be closely proportional to coronal length, but not to height (table

Table 1.26
Dimensions and Wear of M^3 in *Plesiaddax depereti* and *Urmiatherium intermedium*.

	P. depereti	U. intermedium
Mean crown length	28.7	29.9
Mean crown height	29.3	38.0
Hypsodonty index $\left(\dfrac{100 \text{ height}}{\text{length}}\right)$	102	127
Mean annual wear	5.0	5.5
Wear in % of length	17.4	18.4
Wear in % of height	17.1	14.5

1.26). Thus the *Urmiatherium intermedium* tooth will last longer before being worn to the root. (Incidentally, it may be pointed out that, as the tooth wears obliquely, the lingual side will be down to the root when the buccal side may yet be some 5 mm. high.)

In the three *Gazella dorcadoides* populations there is a similar relation between hypsodonty and P.N.L. The animals are equal in size, but the population of Loc. 109 is slightly more hypsodont than the others (table 1.27). Mean annual wear of M_3 is similar in all, but it amounts to a higher percentage of crown height in the less hypsodont populations.

Are the differences within a population great enough to be of selective importance?

In *P. depereti* the standard deviation of the crown height of M^3 in its unworn state is 1.5 mm. Mean annual wear is 4.1 mm., or 3.3 σ. Assume, for instance, that P.N.L. for the animals with mean height of M^3 is 8 years, on an average. Then P.N.L. would be arranged in a distribution with 8 years as a mean and 1 year = 3.3 σ. This gives a P.N.L. of more than 9 years for .1% of the population, between 8 and 9 years for 49.9%, between 7 and 8 years for 49.9%, and less than 7 years for .1%. If the mean is assumed to be 7.5 years, P.N.L. is more than 8 years for 5%, between 7

Table 1.27
Height and Wear of M_3 in three *Gazella dorcadoides* Populations.

	Loc. 109	Loc. 30	Loc. 43
Mean crown height, Age Group (2)	19.2	17.8	18.0
Mean annual wear	4.2	4.2	4.4
Wear in % of height	21.9	23.6	24.4

and 8 years for 90%, and less than 7 years for 5%. Of course coronal dimensions are to some extent correlated with body size, and this, in turn, with rates of wear; it would therefore be somewhat more correct to analyse the case in terms of hypsodonty indices (all dimensions given in percentages of crown length, and standard deviation calculated directly for the index). The mean hypsodonty index is 102, its standard deviation 5.3, and mean annual wear 17.4, or 3.3 σ; the result is identical. If it is assumed that the reproductive period of *P. depereti* and *U. intermedium* began at 3 years of age, and that of *G. dorcadoides* at 2 years, the number of individuals in the three terminal age classes, where this factor assumes importance, will constitute 22, 21 and 20%, respectively, of the potential parents. As the presumably polygenic complex governing hypsodonty in these forms is not known, the problem cannot be posed in genetical terms.

It appears, however, that in a population of quite average variability (V = 5.2 for height of M^3 in *P. depereti*) hypsodonty may have a decided influence on longevity, and hence have selective value. Stirton (1947), after describing the evolution of hypsodonty in horses and beavers, does not consider natural selection on the difference in hypsodonty alone a determining factor. He points out, appropriately I think, that e.g. hardness of dentine and enamel ranks high in importance. But the argument in which he seems inclined to rob hypsodonty of almost all selective value is surely fallacious. Certainly there are other factors that may be as important in survival or longevity; but to conclude that higher tooth crowns could have evolved in continuous sequences through millions of generations, solely due to "genetic linkage with other genes," but without intrapopulation selective value *per se* at any point in time (as Stirton seems to do, though the argument is evasive) is hardly compatible with the extensive possibilities of recombination inherent in sexual reproduction; the conclusion, moreover, is apparently not based on numerical analysis. Pleiotropism of such a convenient kind as to ensure long-range increase of hypsodonty as a by-product of selection for other qualities is obviously absurd. The special case of allomorphosis must also be discounted, as progressive hypsodonty may go with phyletic dwarfing (the *Nannippus* line, see Stirton 1947), and there is really no intrapopulation correlation between hypsodonty and gross size (Simpson 1944). I do not think that progressive hypsodonty can be explained in any terms but selection for hypsodonty. The fact that progressively hypsodont phyla have become extinct has simply no bearing on the question; it just shows that other factors can be of greater importance, at times— not that hypsodonty lacks survival value.

Factors affecting P.N.L. would not seemingly be controlled by selection if mortality rates are so high that most or all individuals die before its termination, or similarly if the individuals become sterile from senility before P.N.L. runs down. It seems, for instance, probable that selection operating on such factors would be slackened in bird populations and in many small mammals. The same would probably be true for the chamois. The mode of life of the latter is obviously a secondary adaptation, and from its more orthodoxly grazing ancestors it has inherited a dental battery the efficiency of which exceeds what is required in its present niche.

AGE AND VARIATION

The classical works of Bumpus (1899), Weldon (1901) and di Cesnola (1907) have established beyond doubt the existence of measurable centripetal selection, affecting the dimensions of animals in such a way as to decrease the variability in older age groups of a cohort. Bumpus (1899, from Huxley, 1942) picked up a number of sparrows (*Passer domesticus*) found helpless in a storm, and found that proximal variants as to wing length survived to a greater extent than did distal ones. Weldon (1901) and di Cesnola (1907) found that inner (younger) whorls of the two land-snails, *Clausilia laminata* and *Helix arbustorum* respectively, were more variable in young individuals than in adult ones; criticism levelled against the significance of the results has been shown by Huxley (1942) not to be pertinent.

Similar data may possibly be obtained from mammalian teeth, in which the coronal dimensions are not affected by growth after eruption; of course, dimensions not affected by wear should be selected. Table 1.28 shows the variation in the dimensions of first molars of *Tapirus i. indicus* Desmarest (from data by Hooijer 1947a). The specimens numbered 20 of which a half still possessed milk teeth and were classed as juveniles; the others were adult, probably of different ages. The means do not show any statistically significant differences. The standard deviations, however, differ markedly; but the results are uncertain on account of the small size of the sample.

Table 1.28
Tapirus indicus Desm. Variation in Dimensions of Upper and Lower First Molars, in Juveniles (with Milk Dentition) and Adults. Computed from Data by Hooijer.

		N	M	σ	V
M^1 anteroposterior dia.	J	10	25.5 ± .32	1.02 ± .23	4.02 ± .90
	A	8	24.6 ± .25	.70 ± .17	2.83 ± .71
M^1 anterotransverse dia.	J	10	26.6 ± .35	1.11 ± .25	4.19 ± .94
	A	9	26.8 ± .26	.79 ± .19	2.93 ± .69
M^1 posterotransverse dia.	J	10	23.7 ± .28	.90 ± .20	3.80 ± .85
	A	9	23.7 ± .27	.82 ± .19	3.45 ± .81
M_1 anteroposterior dia.	J	10	25.6 ± .25	.80 ± .18	3.13 ± .70
	A	8	24.9 ± .21	.60 ± .15	2.41 ± .60
M_1 anterotransverse dia.	J	10	18.5 ± .25	.81 ± .18	4.36 ± .97
	A	10	18.8 ± .24	.75 ± .17	3.98 ± .89
M_1 posterotransverse dia.	J	10	17.7 ± .25	.78 ± .17	4.41 ± .99
	A	10	18.0 ± .25	.77 ± .17	4.31 ± .96

J, juveniles, A, adults.

In the *Ictitherium wongii* population (table 1.29) a similar trend is visible regarding the large functional teeth (premolars), whereas the molars do not exhibit any reduction of variability. The young individuals include the two initial age groups, the remainder being lumped as old individuals. In the phyletic history of the hyaenids, the premolars have "normal" variability throughout, whereas the variability of the molars is steadily increased. These facts invite speculation; but again the samples are too small to permit certain inferences.

Somewhat similar trends were found in the *Ursus arctos* population, whereas that of *Vulpes vulpes* did not show any such difference in variability. The criterion for the division into young and old specimens was the state of wear of the dentition.

In *Ursus arctos,* some differences in average size between teeth of young and old specimens were found. The ratio diagram, fig. 1.32 summarizes the evidence. Males and females were considered separately. In both sexes the old individuals tend on an average to have smaller functional teeth than the young ones; the differences, however, being on the .10 level, are of doubtful significance. The change in average dimensions cannot compare with the marked sexual difference. In the foxes, on the other hand, the trend is precisely the opposite, old specimens tending on an average to

Table 1.29
Ictitherium wongii Zdansky. Variation in Dimensions of Upper Tooth
Crowns, in Young Individuals (with Unworn or Slightly Worn Teeth)
and Older Ones. Original Data.

		N	M	σ	V
Length P^1	Young	18	25.3 ± .31	1.31 ± .22	5.16 ± .86
	Old	12	25.3 ± .27	.94 ± .19	3.73 ± .76
Length P^3	Young	12	16.4 ± .27	.95 ± .19	5.79 ± 1.18
	Old	19	16.5 ± .12	.51 ± .08	3.10 ± .50
Length P^2	Young	8	13.0 ± .16	.44 ± .11	3.40 ± .85
	Old	14	13.2 ± .08	.29 ± .06	2.22 ± .42
Width M^1	Young	14	15.3 ± .16	.59 ± .11	3.86 ± .73
	Old	12	14.8 ± .23	.79 ± .16	5.35 ± 1.00
Width M^2	Young	15	7.1 ± .14	.53 ± .10	7.42 ± 1.35
	Old	11	7.6 ± .15	.50 ± .11	6.60 ± 1.41

have slightly larger teeth than young ones; but here too the significance is, at most, $P = .10$. Evaluation of such trends will necessitate the collection of very large samples.

Nevertheless such studies in conjunction with life table treatment may give valuable data on how selection acts on the phenotypes in nature. Incidentally, there is a point of some importance to the taxonomist. Determination of the "standard range" of variation (Simpson 1941b) of a dimension, based on a sample of mature or old specimens, is liable to error owing to decreased variability on account of centripetal selection. As stressed by Simpson and Roe (1939), values of V < 4 are often dubious and may indicate that the sample is not random.

GROSS SIZE AND RATES OF MORTALITY IN ADULTS

The annual mortality rates (q_x) of the life tables show a fairly great variation from year to year. This is mainly due to the chances of sampling, as the variation is clearly negatively correlated with size of sample. The rates, indeed, show a central tendency in each sample, and some form of weighted mean would apparently give a better estimate of annual mortality in adults of the species in question, than any single observation.

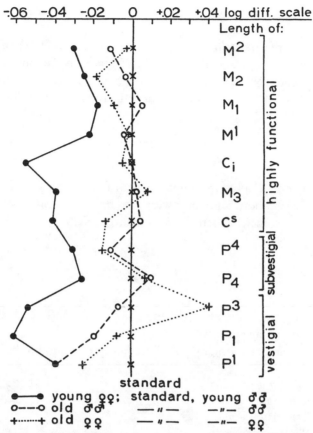

Fig. 1.32.   Ratio diagram comparing coronal dimensions in samples of *Ursus arctos* dentitions. Young females are compared with young males (standard); old males with young males (standard); and old females with young females (standard). Cf. fig. 1.13.

To be sure this rate may actually have changed with increasing age, and certainly did so towards termination of potential natural longevity, but the gradual change in earlier years of life cannot be estimated from the present crude data.

When the mean rate is calculated, the final drop of the l_x curve ought to be left out, since this is mainly caused by other mortality factors than those at work during earlier parts of adult life. The difficulty is to determine where this drop begins; I have arbitrarily chosen the interval in which $100q_x$ increases to 50, which seems to be the most suitable terminal point in the present case.

Averaging the mortality rate is done by fitting a curve of the type $N_z = N_0e^{-kz}$ to the data (Farner 1952), where N_0 is the number alive at the initial date, N_z the number alive at z years after this date, and k equal to ln $(1—M)$. When fitting the curve, I have made use of the formula,

$$(i) \qquad \log S_w = \frac{\Sigma f(d_x d_y)}{\Sigma f(d_x^2)}$$

where d_x is time elapsed from initial date, f number alive at d_x, and $d_y =$ log $f—2$ for a cohort beginning with 100 individuals. $M_w = 1—S_w$. The formula gives somewhat better results than

$$(ii) \qquad \log S_w = \frac{\Sigma (d_x d_y)}{\Sigma (d_x^2)}$$

by increasing the influence of the more abundant, and hence statistically more significant, age groups.

The mean annual mortality rates for adults were found to be:

Gazella dorcadoides	33.5%
Rupicapra rupicapra	24.8%
Ictitherium wongii	20.7%
Plesiaddax depereti	14.3%
Urmiatherium intermedium	10.4%
Ovis d. dalli	8.3%

In this order the species form a series of decreasing rates of mortality. But they also form a series of increasing size.

This trend clearly warrants closer investigation, and I have reconsidered some data on the dynamics of other mammalian populations. Small mammals in particular have been fairly thoroughly studied in this respect (see, e.g., Bourlière, 1951, and references). The rates of survival in deer-mice *(Peromyscus)* seem to average 4 to 6 per cent. (Blair, 1948), the former datum being somewhat more probable, as Blair's population data give a still lower figure (less than 2 percent) when a curve is fitted.

For the snowshoe hare, *Lepus americanus,* Green and Evans (1940) have estimated the mortality during one year to be on the order of 70 percent.

Data on large mammals are in the main highly uncertain. Lundholm (1947) records the ages at death of various subfossil Scandinavian horses. The collection consists partly of domestic Bronze age and later horses, sacrificed at burials, partly again of earlier, wild horses. The age compositions for the two samples are shown in table 1.30.

Theoretically the domestic sample should be subjected to time-specific analysis. There is however an incalculable element of bias in the selection of specimens for sacrifice, and the only inference that can be drawn is that the equality in frequency within the age groups indicates fairly low rates of mortality. The wild population should probably be analyzed dynamically, but the sample is very small, and so the annual mean ($M_w = 7.1$ percent) is not reliable. These data however indicate that rates of mortality in wild horses are low, and comparable to those in large Bovidae.

Plotting of mortality or survival rates against some measure of size may reveal more of the nature of the dependence between these variables, if any such there be. The most appropriate measure of gross size would be net weight or total length. These data are however not known for the fossil forms. A readily obtainable measure is length of skull; relations between head and body size certainly differ, but the variate is a faithful enough indicator of gross size for the present purpose. In fig. 1.33, rates of survival and mortality are plotted against skull length, the latter being scaled logarithmically.

The obvious correlation between size and rate of survival raises the

Table 1.30
Ages of Death of Subfossil Scandinavian Horses

	Frequencies	
Age in Years	Wild	Domestic
0–2	1	2
3–5	2	5
6–6	2	4
9–11	1	5
12–14	1	4
"old"	3	6
totals	10	26

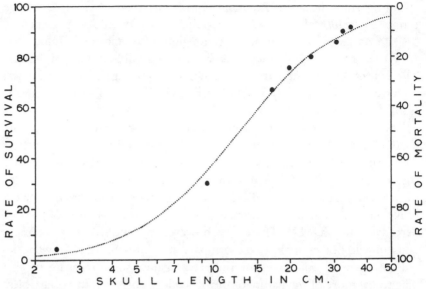

Fig. 1.33   Mean annual rates of survival and mortality for *Peromyscus* spp., *Lepus americanus, Gazella dorcadoides, Rupicapra rupicapra, Ictherium wongii, Plesiaddax depereti, Urmiatherium intermedium,* and *Ovis dalli* (from left to right) plotted against skull length, to indicate relationship between body size and survival. Logistic curve fitted to the data.l

question of the rationalizing of the relationship. Theoretically, mean annual survival cannot be zero (though it may indeed be so in practice), as the curve, $N_z = N_o e^{-kz}$ approaches the horizontal axis asymptotically; neither can it be 100 percent. The required curve should approach o and 100 percent asymptotically, and thus be sigmoid.

One such curve, and the one most commonly used, is the logistic curve. In fig. 1.33, a logistic has been fitted to the data (for the method, see Pearl, 1940). The general form of the equation is for the present purpose

$$S = \frac{100}{1 + Ce^{r(\log x)}}$$

where S is survival and x is skull length. r (which denotes the rate of compounding in the differential equation for geometric increase) was found to be —4.9.

The danger of generalizing from relationships of this kind is justly stressed by Benedict (1938), though in different connexion. But it is certainly equally erroneous to disregard such relationships completely and to deny the reality of manifest trends. Modifications and deviations obviously exist, but the general trend (which is apparently real in the case discussed by Benedict himself) should also be accounted for.

Now in the present case it seems to me rather evident that the logistic curve fitted to the data has no intrinsic meaning at all and is of no use except as a convenient, but perhaps not very accurate, means of description. The constants and terms of this equation cannot be interpreted biologically in a manner similar to that of Gause (1934), who found most suggestive expressions for the struggle for existence, etc., in the equation of the first derivative curve of the logistic.

The descriptive accuracy may be evaluated from Table 1.31, showing estimated and calculated survival, and the differences; the deviations are certainly not great, but it is clear that some asymmetric sigmoid curve might show a better fit. The data are however far from certain enough to warrant such analysis, and for the same reason I deem it superfluous to give the full equation.

Though it is not possible as yet to state the relationship in unequivocal mathematical terms, it seems to be indisputable that adult mortality rates in mammals are to some extent a function of size. But this rule has most important qualifications. It would be a gross error to apply it to mammals

Table 1.31
Actual and Calculated Annual Rates of Survival in Mammals of Different Size.
S_w in Percent.

	actual[a]	calculated[b]	Difference
Peromyscus	4.0	2.83	1.17
Lepus americanus	30.0	34.6	4.6
Gazella dorcadoides	66.5	65.3	1.2
Rupicapra rupicapra	75.2	72.7	2.5
Ictitherium wongii	79.3	79.7	0.4
Plesiaddax depereti	85.7	86.5	0.8
Urmiatherium intermedium	89.6	88.2	1.4
Ovis dalli	91.7	89.6	2.1

[a]Estimates, see the text.
[b]From equation of logistic curve

in general, regardless of affinities and ecological station. The annual mortality rate of the adult lump-nosed bat *(Corynorhinus rafinesquei)* is about 20 percent (Pearson, Koford, and Pearson 1952), or of the same order of magnitude as that of the chamois; the weight of the bat, however, being some 8 grams (ibid.) or about 1/5000 of that of the chamois. Mortality rates in bats are evidently more nearly comparable to those in birds, e.g. the swift, *Micropus apus* (Magnusson and Svärdson 1948), 19 percent.

Data on birds tend to show that no such relation exists between size and rate of survival: "Es sind nicht immer die Grösseren, die länger leben" (v. Haartman 1951). Yet, if the scope is narrowed to a taxonomic unit of the family rank, and to ecologically somewhat similar species, there are data suggestive of a similar relationship. Thus, within the Turdidae, the species arranged in descending order of size (survival data from v. Haartman; original data from Farner on *T. migratorius,* and Lack on the others) exhibit a similar trend:

	SURVIVAL RATE
Turdus merula	60%
T. ericetorum	55%
T. migratorius	48%
Erithacus rubecula	38%

The relationship seems to hold good for small vs. large mammals, as long as the ecological backgrounds do not differ radically, as when bats are compared with other mammals. Its validity for shorter ranges of size is as yet uncertain, and it will probably be found that the picture is complicated by various deviations. It is also possible, and indeed probable, that mammals differing taxonomically and ecologically will be found to constitute distinct series, perhaps parallelling each other, but with different relations between size and survival rates. This is suggested by the conditions in birds (see v. Haartman, 1951), but more data are urgently needed for the clarification of these points.

Finally it should be mentioned (see discussion in Farner 1952) that there is some evidence suggesting that rates of survival may differ in one species from place to place. This also goes to show that rules of this kind are not absolute; which, of course, does not rob them of their value, since they may be useful in the description of average trends and tendencies.

Phyletic Growth

SELECTION FOR LARGER SIZE

Among the empirical "laws" or "principles" of evolution that have been stated by paleontologists, that of phyletic growth (variously referred to as Cope's Law and Depéret's Law) is second in repute only to "Dollo's Law" of the irreversibility of evolution. These two principles, in conjunction, have so saturated much paleontological thought that indeed earlier forms have been excluded from the ancestry of later ones on the single criterion that the earlier one was larger.

The fact that the world is not populated exclusively by giants militates of course against the assumption of phyletic growth as an all-pervading principle, and in many instances (e.g. humming-birds; the fish *Lebistes,* etc.) it is perfectly clear that forms living today are descended from larger ancestors (see Rensch, 1947). In fossil series several instances of phyletic dwarfing are now known (e.g. among the horses), a special case of singular interest being the tendencies to dwarfing in several mammalian species (Primates, Perissodactyla, Carnivora, etc.) during and after the Pleistocene; this has been discussed especially by Hooijer (1947 b, 1948, 1950 a, b, 1952 a) in a series of important contributions.

Like other paleontological "principles" of evolution, that of phyletic growth describes what usually or often happens, rather than what always happens. This holds to some extent even for the most nearly universal principle, that of the irreversibility of evolution, the fact that larger steps in evolution can never be retraced being an aspect of the fact that history never repeats itself (see Simpson, 1950 b, for discussion of this and other principles).

In casting about for the guiding factors of phyletic growth and phyletic dwarfing it seems natural to turn to the relations between gross size and environment that are known to exist in recent species. Bergmann's rule, evidently describing the result of selection in relation to heat-loss, states that homoiothermal forms tend to be larger in colder environments. This adaptation might to some extent account for phyletic growth in mammals toward the close of the Tertiary. It cannot however account for phyletic dwarfing in more recent times, as this process has continued since the *beginning* of the Ice Age.

Another relation is seen in the frequently smaller size of insular species (Rensch, 1924), as compared with their continental relations; this is explained on the assumption of reduced food-supply and *Lebensraum*. But this rule has many exceptions. The largest of all ratites was an insular form.

Probably no all-embracing factor of phyletic dwarfing will be found, and such trends as occur seem in the main to be related to specific adaptations of various kinds. Thus, the recurrent rise of dwarf horses in North America may reflect the occupation of ecological niches like that of the Old World gazelles. However, the *simultaneous* trends described by Hooijer do not seem to be due to such adaptations; they embrace mammals that are not related taxonomically and evidently not always ecologically. The one similarity that can be found seems to be that the species exhibiting this trend are large mammals. The factors may possibly be identical with those causing the wholesale extinction of many giant mammals at the same time; this is an important problem, and one that has defied all attempts to clarification so far.

We cannot expect either to find one single factor of phyletic growth. But there are obviously some selective factors operating in this direction, regardless, at times, of specific adaptive value. This may account for the fact that phyletic growth seems to be more common than phyletic dwarfing, and thus has come to be regarded as an empirical principle.

The selection pressure arising from the higher fertility of larger poikilothermous forms has been discussed above. It may here be added, that such conditions seem to obtain in some mammals as well; at least intraspecifically. Venge (1950) shows that larger rabbits are on an average more fertile than smaller ones. This relation, however, does not have general validity; on the contrary, small mammals tend on an average to be more fertile than large ones.

The decrease of annual mortality with increase of gross size, however, is evidently a potential selective factor in mammals. This is implicitly stated by Simpson (1944) in his discussion of phyletic growth. It is also advanced by Newell (1949) to account of the same phenomenon in invertebrates. Both ascribe the trend to the probable survival advantage of larger bulk. The evidence discussed above indicates that this may be true for mammals; as to the lower vertebrates and invertebrates, it remains yet to be proved. Comparative studies of mortality in related species differing in size, and of changes in mortality with growth in a cohort, are required to give the answer.

These are not, however, the only ways in which phyletic growth may affect the survival and population dynamics of a species. In the following pages, some other aspects of the problem will be discussed, with special reference to the mammals.

GROSS SIZE AND POTENTIAL LONGEVITY

Rensch (1947) states as a general rule, "Ganz allgemein werden grössere Warmblüter älter." To Rubner (1908), we owe the discovery that the total amount of heat produced per weight unit during life is closely similar in mammals of very different sizes. The length of life is inversely related to "rate of living" (Pearl, e.g., 1940). Backman (1943) has developed the idea still further, into the concept of *organic time*—with most interesting philosophical connotations.

Rubner also recognized that his "law" was not of general validity for the mammals; for instance, man enjoys by far a longer span of life than seemingly warranted by his body size.

Evidently it is possible to state the relation between gross size and longevity in terms of logarithmic regression. Rubner (1908) complained of the paucity of data on potential longevity; since then, especially by the untiring efforts of Major Flower, on whose great compilation (1931) I have drawn, many data of this sort, concerning animals in captivity, have been gathered.

Potential longevity—the maximum span of life of a species under optimal conditions—may certainly not be equated with longevity in captivity, and the bias will be somewhat inconsistent, as some animals thrive better in captivity than others. These data are however the best ones available at the moment, and they may be used for the present purpose.

Potential longevity (here and subsequently *in captivity,* unless otherwise stated) is a variate and probably fluctuates around a mean; and the localization of this mean is more important, for the present purpose, than the stating of maximum records. The latter ones will tend to deviate from the mean, the deviation increasing with the number of observations, and hence will not do for purposes of comparison. The sample from which the mean is to be computed must be adequate; the individuals should have died from senility, not by accident, disease or maltreatment. Unfortunately these prerequisites cannot be met at pres-

Table 1.32
Longevity in the Cat Family

Species	No. of Observations	Mean Longevity (Years)
Panthera leo	24	19
P. onca	15	14
P. tigris	13	16
P. pardus	13	15.5
Acinonyx jubatus	3	14.4
Felis concolor	6	12.6
F. rufus	5	12.7
F. caracal	5	12.2
F. chaus	3	9.5
F. canadensis	3	12.2
F. lynx	2	13.5
F. bengalensis	2	11.0
F. viverrina	2	9.9

ent, and the only available solution is to compute the arithmetic means of the ages at death of animals with long records in captivity. The results arrived at in this way are tentative in nature, but they are sufficient for the demonstration of trends and relationships, as will be shown below.

For the present purpose it would seem most adequate to select a fairly uniform taxonomic group, yet with appreciable variation in size. The cats fulfill these conditions (one living subfamily with but three closely related genera, and great size range). Computation from Flower's data gave the following results: shown in table 1.32.

The data are plotted against body length (excluding tail) in fig. 1.34; data on length are taken from modern editions of Brehm and other reference works. A linear pattern is clearly recognizable; so clearly that, in my opinion, it lays the burden of proof on him who does not consider the relation valid. The relation is expressed by the equation,

$$y = bx^k,$$

and in the present case, $k = .49$. The deviations may well be attributed to the inconsistent sources of bias briefly reviewed above, and to small-sample effects, but it would of course be premature to conclude that these are the only factors involved. It can however be inferred that the Felidae form a "series" in which potential longevity is, at least mainly, a function of size.

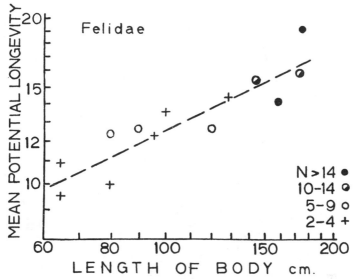

Fig. 1.34. Average potential longevity plotted against body length for 13 felid species.

The regressions for other groups of carnivores have been obtained in a similar way. Fig. 1.35 shows the results, concerning the Ursidae, Mustelidae, and Viverridae + Hyaenidae (considered as one group), as compared with the Felidae. Furthermore, the regression for an artiodactyl family, the Bovidae, has been computed. The coefficient k varies as follows:

Taxonomic Group	k
Bovidae	.40
Mustelidae	.48
Felidae	.49
Viverridae + Hyaenidae	.69
Ursidae	.73

The differences between the various carnivore regressions are not great (the size range in the Ursidae, with the highest value of k, is rather short), and it is possible that one single curve may describe this relation fairly adequately for the majority of living fissipedes. Again, it may be found that each family, or perhaps each genus, or even smaller unit, is characterized by a peculiar relationship between size and longevity; if so, these relations should not however differ sharply within the fissipedes.

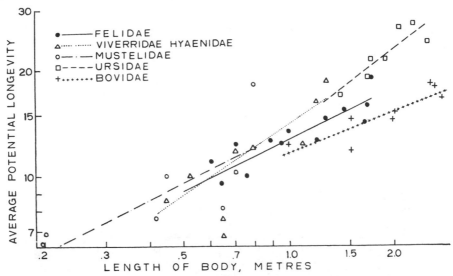

Fig. 1.35. Regressions of average potential longevity on body length for taxonomic units, as labelled.

For the Bovidae, the coefficient *k* is lower than for the carnivores, and this difference is probably valid. It would seem that at least the major taxonomic units differ markedly as regarding this relation; there can be no doubt that the regression for man, and probably for other Primates as well, is very different from those for carnivores and artiodactyls.

Thus phyletic growth can be demonstrated to affect the survival of a cohort in two ways: by decreasing the annual rates of mortality, and by increasing the potential longevity. Under certain circumstances these factors may conceivably operate in perfect harmony; then, provided that other factors are equal, the *form* of the survivorship curve would not change. The mean longevity of the cohort would be raised, but the curves, when superposed on the basis of deviation from mean longevity, would be congruent, as is almost the case with the curves for *Ovis dalli* and man in fig. 1.26.

By increasing the absolute area under the survivorship curve—or, in other words, by increasing the potential opportunities for reproduction of the cohort—size increase is clearly of selective advantage, and the present factor is probably one of the important constituents of phyletic growth. As with those previously discussed, it is probably checked by specific adaptive needs related to the ecological niche of the population. Competition from the population occupying the "next larger" niche may account for part of the counterselection. When the "next larger" niche is unoccupied, however, the various factors of selection pressure favouring larger size may be effective in the long run.

GROSS SIZE AND DEVELOPMENT

The length of the gestation period is influenced by several factors. Basically it is related to the problem of growth rates. This is apparent e.g. when comparison is made between the crosses *Equus caballus* ♂ × *E. asinus* ♀ and *E. asinus* ♂ × *E. caballus* ♀ (see Asdell 1946). In the former case, the gestation period is reduced from the normal span in *E. asinus,* possibly because the foetus attains critical size at an earlier stage in development; in the latter case, gestation is prolonged in excess of the normal span in *E. caballus,* perhaps for the opposite reason. Maternal influence may also be involved (Asdell 1946). Growth rates, in turn, are evidently related to rates of living and thus part of the complex touched upon in the previous section. On an average, the rates of development, like the rates of living, are lower in large mammals than in small ones; hence the period of gestation will average longer in the larger mammal. This relation may obviously be stated in numerical terms.

Some additional considerations are necessary, though an exhaustive discussion of the very complex problems is outside the scope of the present paper. In different groups of mammals, the young are born in greatly varying stages of development. Carnivore cubs are generally quite helpless at birth, often blind, and need a long period of maternal care. Most ungulates, on the other hand, are well-developed at birth and fairly able to take elementary care of themselves within a short time, doubtless an adaptational character. In marsupials the young are born in an embryonic stage and are reared postnatally for an extended period in the pouch. Thus taxonomically diverse forms cannot be directly compared. Even in closely related forms minor differences of the same type are found. Thus the gestation of the mare averages 335 days, and that for the ass 365 days (Asdell 1946). This may be correlated with a slightly more advanced stage of development in the newborn ass, but it may also reflect lower rates of development and living (the data of Flower,

1931, seem to suggest that potential longevity in asses exceeds that of many larger Equidae).

Delayed implantation, especially common in the Mustelidae, is another phenomenon causing deviations from the normal duration of pregnancy.

In fig. 1.36 length of gestation is plotted against body length for the Felidae (data on gestation from Asdell 1946 and Bourlière 1951). Here, again, an essentially linear pattern emerges. Deviations are however apparent within the subgenus *Felis (Lynx),* and these are probably real. In general, however, the relation between size and gestation is similar to that between size and longevity; also to the extent that the regressions vary for different taxonomic units (see fig. 1.37).

LIMITS TO PHYLETIC GROWTH

Increase of body size in mammals, as has been shown, is normally accompanied by extended potential longevity, increased length of gestation period, and higher rate of survival. Other factors being equal, the result will be selection for larger size, and, if suitable mutations are produced and the selection pressure is strong enough to override counterselection, phyletic growth.

Such size increase cannot, of course, go on *ad infinitum;* in land mammals there are definite limits to phyletic growth. Some of them, e.g.

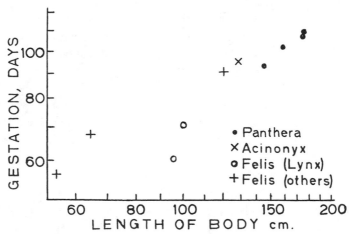

Fig. 1.36.   Mean length of gestation period in Feliadae, plotted against body size.

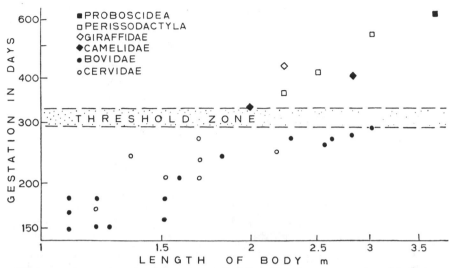

Fig. 1.37. Mean length of gestation period in various large mammals as labelled, plotted against length of body, to indicate deflection of regression for Bovidae immediately beneath the threshold zone.

the mechanical limits, are ably discussed e.g. by Huxley (1942) and need not be stressed further. The existence of a physiological limit is suggested by the diminishing of the surface-bulk ratio that may ultimately lead to inability to dispose of the heat produced by the basal metabolic processes (see Benedict, 1938); seemingly, the giant *Baluchitherium* was not far from this limit.

When considering the breeding cycles in middling to large species of mammals, another complication is clearly seen to arise from the relationship between gross size and gestation period. Most larger mammals breed during a fixed and relatively short season of the year, and parturition takes place at another season, making it possible for the females to take care of their young for some time before breeding anew. With excessive size increase, however, the interval between parturition and renewed breeding may at last approach zero; and finally the gestation period may span the whole year or even a longer time. This is likely to occur with large herbivores only, but here the complication is probably of no mean consequence.

With the approach to this limit in a seasonally breeding population, the direction of selection will seemingly be reversed, for now there is a definite disadvantage in further increase of size. The larger females will tend to have a longer gestation period, and may not have sufficient time to recover from parturition and to develop the physiological and ethological conditions required for fertilization; thus, their breeding may be belated, and the next year they may lag after the time table so much that they may not be able to find a mate at all. Moreover, the young will be born more and more out of season, with increased juvenile mortality as a result. In this way the number of offspring of the larger individuals would be reduced, and the population should attain stability at "ceiling size."

This state of matter is seen in a very large number of large herbivores, as for instance domestic cattle and most large bovids. On the other hand, the species in which the gestation period is too long to permit annual parturition per female are relatively few, such as recent Equidae, Rhinocerotidae, Tapiridae, Camelidae, Giraffidae, Proboscidea. These forms have somehow managed to traverse the "threshold."

That there is really such a threshold effect can be easily demonstrated if length of gestation period is plotted against body size (fig. 1.37). It is seen that the data on length of gestation tend to crowd slightly below the threshold duration in large Bovidae, to some extent independently of size. This shows that selection may favour a change in the relation between gross size and length of gestation period, if there is sufficient advantage in large size. But this process cannot be pushed very far, for the result will be some under-development in the newborn, lowering the level of adaptation and leading to higher mortality in the young. The well-developed state of the newborn in herbivores is an important adaptation, in relation to feeding habits as well as to enemies.

On the other hand, parturition might cease to be limited to one particular season of the year. This is evidently not realized in many large mammals, except elephants. Seasonal parturition seems to be much more common among large mammals, for instance the giraffes, camelids, horses, and evidently the rhinoceroses (see also *Chilotherium*, above), all of which are larger than "threshold size." This is doubtless a feature of adaptation. It does not preclude the possibility that the ancestors of these forms, during the part of their evolution when the size threshold was traversed, bred at any time of the year, but it renders this less probable.

A possibility that may well be the explanation in many cases would be somewhat similar to the quantum shift of Simpson (1944). The transition from breeding interval 1 year to breeding interval 2 years would not, of course, occur in one saltation. In the transitional populations, the interval would vary, at the beginning, from one individual to another and in the same individual from time to time, the mean, however, being somewhere between 1 and 2 years and moving in either direction.

The initial result from excess of "ceiling size" would seemingly be a decrease of numbers within the population. "A change in the reproductive output of a species may have a considerable effect on its density, as in the case of the short-eared owl, cited by Elton (1927)" (Solomon 1949). Under such circumstances it would seem that the density-dependent mortality factors would slacken their grip, and the population might attain stability after a reduction in number.

Such a contraction would tend to split the population, at least in marginal zones, into "islands" (see e.g. Timofeeff-Ressovsky 1939), inhabiting optimal biota, with unpopulated spaces in between, and limited intermigration. This is, according to Wright (e.g. 1945), the optimal condition for rapid evolution. It is to be assumed that the majority of these small populations may become extinct, or revert to the initial stage as regarding size; a minority, however, might effect the transition. The main characteristics of these individuals would be large size and raised mean longevity. Such populations would muster the incipient giants of the animal world.

THE SIZE THRESHOLD AND THE FOSSIL RECORD

We have set up a theoretical case for quantum evolution. It remains to be seen whether any actual phyletic sequence shows peculiarities typical of such an occurrence (gap in record or decrease of number of found specimens for some period, coinciding with fairly rapid size increase).

One's mind inevitably turns to that inexhaustible textbook example, the horse sequence. This has been cited—incorrectly more often than not—as evidence for practically every evolutionary principle that has ever been coined. This makes for some reluctance on the part of an author whose knowledge of this evolutionary record stems, in the main, from the literature. I have consulted, *inter alia,* Stirton (1940, 1947), Matthew (1926), Osborn (1918), Simpson (1944, 1951), Romer (1947).

The series up to *Merychippus*—disregarding early giant sidebranches—is distinctly below the threshold. From *Merychippus* arose the dwarf lines *(Nannippus* and *Calippus);* the *Hipparion—Neohipparion* groups; and the *Pliohippus-Equus* and *Pliohippus-Hippidion* sequences. *Hipparion* and *Neohipparion,* though larger than their ancestors, were seemingly still below the threshold. It would seem highly probable that the transition occurred within the scope of the genus *Pliohippus.*

There is, within this genus, a fairly rapid size increase; the transitional populations are, however, rather poorly documented, at least in comparison with earlier and later parts of the series. The whole aspect is that of a minor case of the "systematic deficiences" of Simpson (1944). Moreover, another major change took place in *Pliohippus,* the reduction of the lateral toes; and this, similarly, is so poorly documented and seemingly so rapid that Robb (1936) discusses the possibility of its being due to one single mutation.

The paleotheres may possibly have parallelled *Pliohippus* in this respect. But it is possible that these considerations do not apply to the early Tertiary mammalian giants. Their gestation period may well have been markedly shorter than that in modern mammals of equal size; it is not unlikely that the long period of gestation, and the well-developed state of the newborn, is an adaptation peculiar to modern herbivores. The profusion of archaic mammalian giants seems to suggest that there were no inhibitions of this kind to their phyletic growth. Their deficiency in intelligence, their presumably lower standard of postnatal care, and the high juvenile death rates resulting from such factors and from the helpless state of the newborn young may have been balanced by high fecundity and relatively short breeding interval.

Finds of seasonally deposited early Tertiary fossils would be of great interest. Such fossil occurrences probably exist, and collections may actually have been made, though without the fact being realized. In the research on population problems a restudy of the great fossil collections may play an important part.

The appearance of gigantism in lower vertebrates was presumably not subject to inhibitions of the kind here discussed. We know as yet very little of the population dynamics of the dinosaurs. The large sauropods, when fully grown, were in all probability largely immune to predatory attacks. Their temperature regulation may be estimated, from the experiments of Colbert et al. (1946) to have been very efficient and for all practical purposes approaching the homoiothermal state. The fact that these creatures got along with an extremely feeble dentition would seem to indicate that there was an abundant and easily obtainable supply of food. On the other hand, the fecundity of the dinosaurs seems to have been high. The clutches of *Protoceratops* number 30 to 35 eggs (Brown and Schlaikjer 1940). If the seemingly long-lived dinosaur adults bred once per annum, a very high juvenile mortality seems to have been a characteristic feature of the populations. Study of dinosaurian population dynamics—the excellent *Protoceratops* collection would certainly be worth a restudy along these lines—might conceivably throw some light also on the problem of dinosaur extinction; this is still a moot question, fully as baffling as the late Pleistocene extinction of many large mammals (see Colbert 1949).

In this discussion, some ways in which the problem of gigantism may be tackled, from the population dynamics point of view, have been touched upon. But the giant forms, their often sudden rise and speedy extinction, will for long remain a spectacular problem to the evolutionist.

Conclusions and Desiderata

The perhaps most important conclusion of Part II is that study of fossil population dynamics is feasible and that samples lending themselves to such research are by no means so rare as might be thought. The crucial point is the determination of individual age. There are two possible solutions to this problem. The fossils may in themselves indicate the age of the animal, as when hard structures have grown discontinuously; or else the sample may have been deposited seasonally, whereby continuous ageing or growing processes appear split into groups one year apart in age; on the assumption, of course, that reproduction is seasonal.

The present study of the age composition of fossil mammalian populations depended on the facts that deposition was seasonal and wear of dentition rapid enough to create detectable changes in the course of a year. Similar methods might profitably be applied to the *Merychippus* sample of Matthew (1924) and the *Stenomylus* sample of Loomis (1910), for instance. On the other hand, samples of lower vertebrates have also been found to indicate seasonal deposition (Olson and Miller 1951), and similar evidence will probably be found in invertebrates as well. Such forms, growing throughout life, will exhibit age groups, and life table analysis may be possible.

It seems probable, however, that animals in which the fossil structures *per se* give evidence of individual age will play a more important role in the study of population dynamics. Absolute or relative measures of age are available for many invertebrate groups and even for a few mammals; again, in some instances, estimates may be formed regarding fossils or subfossils when reliable living standards are known.

It should however be pointed out that in many instances the animal changes its biotope during its life history, and so forms in different stages of development may be predominately found in different types of facies. Such factors require careful weighing of the evidence, and it may be that segments only of the survivorship curves may be obtained for many forms (as for the ostracod, *Beyrichia jonesi;* see above).

There are several points on which more information is badly needed. In the first place, the major neoecological results pertain to avian populations; paleontological study may yield much-needed data on other groups: mammals, lower vertebrates, and in particular invertebrates. In this way more information may be gained also on the types of survivorship curves that occur in nature, and their relation to the structure and the ecological niche of the animals.

The problem of juvenile mortality is one of the uncertain points in neoecology. So far, rates of juvenile mortality have only been inferred, not observed. Unfortunately this is a problem where paleontology cannot be of great help. In most cases, juvenile or larval stages do not fossilize to the same extent that adult ones do, and the result will generally be under-representation or complete absence of juveniles. In some rare instances unbiassed juvenile age groups may be present; this seems at least possible regarding the *Protoceratops* sample.

In paleontological work it seems natural to stress the evolutionary aspect. Here there are two possible points of attack. In the first place, the selective factors contributing to typical evolutionary processes may be evaluated: the dependence of survival and mortality rates on size; the factors involved in progressive hypsodonty, and so on. On the other hand, there is the possibility of studying the actual phylogeny of population dynamics within successions of evolving populations. This would seem to be of particular interest when it comes to extinct groups, like trilobites and ammonites, and might contribute to the problem of their disappearance.

These are then some of the potentialities of fossil population dynamics. The seeming paradox of calculating the expectation of life for an animal that died perhaps millions of years ago resolves itself into a useful method for the study of evolution. The paleontologist cannot coax Father time into breathing new life into his fossils; but there is no reason why he should reject the tools that are at his disposal. Should the present paper stimulate the ecological study of fossil populations, its main purpose will be fulfilled.

SUMMARY

Part I.

The correlation between tooth crown dimensions has been studied in a number of mammal populations. The rule of neighborhood holds good in general, adjacent teeth being more closely correlated as to size than alternate or more distant teeth. This is explained on the assumption of genetically constituted "correlation fields" governed by polygenic complexes, in which the major effect of the single factors is exerted within a limited part of the range. These fields appear to be identical in both upper and lower tooth rows, since antagonistic (occluding) teeth showed the highest correlation of all, of the same order as that between homologous left and right teeth, where genetic identity may be safely presumed. The fields may be represented by contour line diagrams showing values of the transformed correlation coefficient z. With the rudimentation of teeth the corresponding correlation tends to be weakened. In the Hyaenidae, the evolutionary of the correlation field in connexion with the rudimentation of the upper molars may be traced.

The correlation tends on an average to be somewhat higher in males. In some instances a positive correlation between size of skull and the size of functional teeth is found.

Similar correlations exist within other serial structures: in the plumage of the bird wing, the antennae of insects, and the digits of tetrapods. A peculiar feature is the raised correlation between end members of a series, which is found in some instances.

The status of the Pontian Chinese hyaenid populations is considered in this connexion. *Ictitherium sinense* Zdansky is regarded as synonymous with *I. gaudryi* Zdansky. From a discussion of the variation of vestigial teeth it is inferred that ?*Lycyaena dubia* Zdansky may probably be conspecific with *I. hyaenoides* Zdansky. The main features of early hyaenid evolution are discussed, with observations on some evolutionary principles. An aberrant specimen that may possibly be a true *Lycyaena* is compared with other populations, and the possibility of hybridization discussed.

Part II.

The age grouping in the Lagrelius collection from the Chinese *dorcadoides* fauna evidently results from seasonal mass deaths and subsequent deposition, the developmental stages in each species clearly being one year apart in age. Juvenile stages are separable on the basis of tooth development in many species; in some, moreover, the adult age groups could be delimited on the basis of the wear of the dentition. The remains are probably of minor populations killed by abiotic factors, and various data suggest that floods are probably responsible for the mass deaths.

The data have been treated in life tables by time-specific analysis. Comparison is made with the population dynamics of larger recent mammals, and instances of "diagonal" and "convex" survivorship discussed.

Data on juvenile mortality are uncertain, the sampling being biassed in favor of adults, but synthetic treatment indicates that losses of about 7/10 up to puberty are common among mammals.

In general selection would seem to be more effective in earlier stages, where the cohort is strong in numbers. Particular aspects of selection are related to hypsodonty and gross size. It is found that intrapopulation variation in hypsodonty may be of selective value. Gross size seems on an average to be correlated with rates of mortality in terrestrial mammals; when mean annual survival is plotted against a measure of size, the data tend to fall approximately along a logistic curve. The rule, however, has important qualifications.

It may however be one of the important factors in phyletic growth. Another factor is the positive relation between gross size and potential longevity, and, in invertebrates and coldblooded vertebrates at least, the higher fecundity of larger forms.

In mammals there is also a relation between size and length of gestation, which, in large herbivores, may put a temporary limit to phyletic growth. "Threshold size" would seem to be attained when the length of gestation approaches 1 year in seasonally breeding mammals. Some evidence in the evolution of the horse suggests that the threshold may be traversed by a quantum shift.

The study of ecology and evolution may profit from research on "fossil population dynamics." Large samples of invertebrates may be particularly well suited to this kind of treatment; a life table is given for a Silurian ostracod, where age is given in terms of moult stages.

NOTES

1. This is not, of course, necessarily true.

2. This is not intended as a statement that geographic speciation is the only way of evolution. Strictly temporal clines, with continuous change in situ, are known (e.g., Brinkmann, 1929). But evolution in conjunction with migration seems to be one of the major modes. This is not invalidated by the distinct possibility that, in Kaufmann's actual example (late Cambrian Olenidae), the evidence may be misinterpreted; the diagnostic differences in relative breadth of the small trilobites may really be due to deformation from pressure, correlated with lithological conditions, as Dr. Jaanusson *(in verbis)* kindly informs me.

3. Such size increase may in part be due to random elimination of genes in migrating populations (Reinig 1938, 1939), but would seem generally to be conditioned by selection (Rensch 1938, 1939, 1947).

4. The ratio diagram was developed by Simpson (1941a) and is now in wide use. Logarithms of dimensions are compared with one population as a standard (log diff. 0, straight line). Full explanations are given in the cited work.

5. It is perhaps superfluous to belabor a too rigid application of Dollo's law by pointing out that the subsequent newly formed correlation between M^1 and M^2 (in *I. hyaenoides*) is an instance of reversal in evolution.

6. Study of the number of corpora lutea, as in Wheeler's (1934) treatise on the fin whale, is of course impossible in fossil material.

7. A brief description of the age groups and life tables of *P. depereti, Urmiatherium intermedium, Gazella dorcadoides,* and *Ictitherium wongii,* is given in Kurtén (1953).

8. The period may nevertheless have been long. Recent secular fluctuations in the composition of faunas are correlated with climatic changes (see Kalela 1949, 1950), but these changes are evidently much more violent in our time than before the Pleistocene: we still live in an Ice Age.

9. For starved *Drosophila,* of course, the distribution of resistance to starving should be substituted for the distribution of senescence.

REFERENCES

Abel, O. 1912. *Grundzüge der Palaeobiologie der wirbeltiere.* Stuttgart.

Abel, O. 1927. *Lebensbilder aus der Tierwelt der Vorzeit.* Jena.

Aller, W. C., A. E. Emerson, O. Park, Th. Park, and K. P. Schmidt. 1949. *Principles of Animal Ecology.* Philadelphia: W. B Saunders.

Alpatov, W. W. and A. M. Boschko-Stepanenko. 1928. Variation and correlation in serially situated organs in insects, fishes and birds. *Amer. Naturalist 62.*

Asdell, S. A. 1946. Patterns of mammalian reproduction, Ithaca: Comstock.

Backman, G. 1943. *Wachstum und organische Zeit.* Leipzig.

Bateman, W. 1892. *Materials for the Study of Variation, Treated with Special Regard to Discontinuity in the Origin of Species.* London.

Benedict, F. G. 1938. Vital energetics. A study in comparative basal metabolism. *Carnegie Inst. Publ.* 503.

Bergmann, C. 1847. *Uber die Verhältnisse der Warmeoekonomie der Thiere zu ihrer Grosse.* Götting. Studien I.

Blair, W. F. 1948. Population density, life span, and mortality rates of small mammals in the blue-grass meadow and blue-grass field associations of southern Michigan. *Amer. Midl. Naturalist* 40.

Bodenheimer, F. S. 1938. *Problems of Animal Ecology.* London.

Bohlin, B. 1926. Die Familie Giraffidae. *Pal. Sinica.* C 4:1.

Bohlin, B. 1935. Cavicornier der Hipparion-Fauna Nord-Chinas. *Pal. Sinica* C 9:4.

Bohlin, B. 1939. *Gazella (Protetraceros) gaudryi* (Schl.) and *Gazella dorcadoides* Schl. *Bull. Geol Inst. Upsala,* 28.

Bourliere, F. 1951. *Vie et moeurs des mammiferes.* Payot, Paris.

Brehm, A. E. 1929. *Djurens liv.* Ed. H. Rendahl, Malmö.

Brehm, A. E. 1938. *Djurens liv.* Ed. S. Ekman. Stockholm.

Brinkmann, R. 1929. Statistisch-biostratigraphische Untersuchungen an mitteljurassischen Ammoniten über Artbegriff und Stammesentwicklung. Abh. Gesellsch. Wiss. Göttingen. *M.-P. Kl. N. F.* 13:3.

Brown, B. and E. M. Schlaikjer. 1940. The structure and relationships of *Protoceratops. Ann. New York Acad. Sci.* 40:3.

Bumpus, H. C. 1899. The elimination of the unfit as illustrated by the introduced sparrow, etc. Bird Lecture Woods Hole, 1898.

Butler, P. M. 1937. Studies of the mammalian dentition, I. The teeth of *Centetes ecaudatus* and its allies. *Proc. Sool Soc. London,* Ser. B. 107.

Butler, P. M. 1939. Studies of the mammalian dentition; differentiation of the post-canine dentition. *Proc. Zool. Soc. London,* Ser. B. 109.

Cesnola, A. P. di 1907. A first study of natural selection in *Helix arbustorum. Biometrika* 5.

Colbert, E. H. 1933. Siwalik mammals in the American Museum of Natural History. *Trans. Amer. Phil. Soc.* New Ser. 26.

Colbert, E. H. 1939. Carnivora of the Tung Gur formation of Mongolia. *Bull. Amer. Mus. Nat. Hist.* 74.

Colbert, E. H. 1949. *The Long Reign of the Dinosaurs.* Northwest Missouri State Teachers Coll. Studies, 1949.

Colbert, E. H., R. B. Cowles and C. M. Bogert 1946. Temperature tolerances in the American alligator and their bearing on the habits, evolution, and extinction of the dinosaurs. *Bull. Amer. Mus. Nat. Hist.* 86.

Colbert, E. H. and N. D. Newell, 1948. Paleontologist—biologist or geologist? *Jour. of Paleont.* 22:2.

Couturier, M. 1938. *Le chamois.* Grenoble.

Deevey, E. S. Jr. 1947. Life tables for natural populations of animals. *Quart. Rev. Biol.* 22:4.

Dobzhansky, Th. 1944. On species and races of living and fossil man. *Am. Jour. Phys. Anthrop.* New Ser. 2:3.

Ellenberger, W. and H. Baum 1926. *Handbuch der vergleichenden Anatomie der Haustiere, 16 Aufl.* Berlin.

Elton, C. 1927. *Animal Ecology.* London.

Errington, P. L. 1946. Predation and vertebrate populations. *Quart. Rev. Biol.* 21.

Farner, D. S. 1952. The use of banding data in the study of certain aspects of the dynamics and structures of avian populations. *Northwest Science* 26.

Flower, S. S. 1931. Contributions to our knowledge of the duration of life in vertebrate animals. V. Mammals. *Proc. Zool. Soc.* London.

Gaudry, A. 1862–67. Animaux fossiles et Geologie de l'Attique.

Gause, G. F. 1934. *The Struggle for Existence.* Baltimore, Williams and Wilkins.

Gause, G. F., O. K. Nastukova and W. W. Alpatov 1934. The influence of biologically conditioned media on the growth of a mixed population of *Paramecium caudatum* and *P. aurelia Jour. Anim. Ecol.* 3.

Green, R. G. and C. A. Evans 1940. Studies on a population cycle of snowshoe hares on the Lake Alexander area . . . *Jour. of Wildlife Management,* 4.

Gregory, W. K. and M. Hellman 1939. On the evolution and major classification of the civets (Viverridae) and allied fossil and recent Carnivora. *Proc. Amer. Phil. Soc.* 81.

Haartman, L. von 1951. Der Trauerfliegenschnäpper. II. Populationsprobleme. *Acta Zool.* Fennica, 67.

Hickey, J. J. 1952. Survival studies of banded birds. Special Scientific Report (Wildlife) No. 15, Washington.

Hoglund, H. 1943. On the biology and larval development of *Leander squilla* (L;) forma *typica* de Man.

Hooijer, D. A. 1947a. On fossil and prehistoric remains of *Tapirus* from Java, Sumatra and China. *Zool. Meded.* 27, Leiden.

Hooijer, D. A. 1947b. Pleistocene remains of *Panthera tigris* (Linnaeus) subspecies from Wanhsien, Szechwan, China, compared with fossil and recent tigers from other localities. *Amer. Mus.* Nov. 1346.

Hooijer, D. A. 1948. Prehistoric teeth of man and of the orang-utan from Central Sumatra, with notes on the fossil orang-utan from Java and southern China. *Zool. Meded.* 29, Leiden.

Hooijer, D. A. 1950a. The fossil Hippopotamidae of Asia, with notes on the recent species. *Zool. Verh.,* p. 8. Leiden.

Hooijer, D. A. 1950b. The study of subspecific advance in the Quaternary. *Evolution* 4.

Hooijer, D. A. 1952a. Some remarks on the subspecies of *Phalanger ursinus* (Temminck) and of *Lenomys meyeri* (Jentink) from Celebes. *Zool. Meded.* 31, Leiden.

Hooijer, D. A. 1952b. Australomelanesian migrations once more. *Southwestern Jour. Anthropology* 8:4.

Huxley, J. 1932. *Problems of Relative Growth.* London.

Huxley, J. 1939. Clines: an auxiliary method in taxonomy. *Bijdr. Dierk.* 27.

Huxley, J. 1942. *Evolution. The Modern Synthesis.* London.

Kalela, O. 1949. Changes in the geographic ranges in the avifauna of northern and central Europe in relation to recent changes in climate. *Bird-Banding* 20.

Kalela, O. 1950. Zur sakularen Rhythmik der Arealveranderungen europaischer Vogel und Saugetiere, mit besondere Berucksichtigung der Uberwinterungsverhaltnisse als Kausalfaktor. *Ornis Fennica* 27.

Kaufmann, R. 1933. Variationsstatistische Untersuchungen uber die Artabwandlung und Artumbildung an der oberkambrischen Trilobitengattung *Olenus* Dalm. Abh. *Geol. Palaont. Inst. Greifswald* 10.

Kurtén, B. 1952. The Chinese *Hipparion* fauna. A quantitative survey, with comments on the ecology of the machairodonts and hyaenids and the taxonomy of the gazelles. *Soc. Sci. Fennica, Comm. Biol.* 13:4.

Kurtén, B. 1953. Age groups in fossil mammals. A preliminary report. *Soc. Sci. Fennica, Comm. Biol.* 13:13.

Lewentz, M. A. and M. A. Whiteley 1902. A second study on the variability and correlation of the hand. *Biometrika* 1.

Loomis, F. B. 1910. Osteology and affinities of the genus *Stenomylus. Amer. Jour. Sci.* 29.

Lundholm, B. 1947. Abstammung und Domestikation des Hauspferdes. *Zool. Bidr. Uppsala,* 27.

Macdonnell, W. R. 1913. On the expectation of life in ancient Rome, and in the provinces of Hispania and Lusitania and Africa. *Biometrika,* 9.

Magnusson, M. and G. Svardson 1948. Livslängd hos tornsvalor *(Micropus apus* L.). *Vår Fågelvärld* 7:4.

Martin, P. 1914–19. *Lehrbuch der Anatomie der Haustiere,* vols. 2-3. Stuttgart.

Matthew, W. D. 1924. Third contribution to the Snake Creek fauna. *Bull. Amer. Mus. Nat. Hist.* 50.

Matthew, W. D. 1926. The evolution of the horse: A record and its interpretation. *Quart. Rev. Biol.* 1.

Mayr, E. 1944. *Systematics and the Origin of Species.* New York.

Merriam, J. C. 1908. Death trap of the ages. In Published papers and addresses of John C. Merriam. Carnegie Inst. Publ.

Merriam, J. C. 1909. A death-trap which antedates Adam and Eve. Ibid.

Merriam, J. C. and C. Stock 1925. Relationships and structure of the short-faced bear, *Arctotherium,* from the Pleistocene of California. Ibid.

Merriam, J. C. and C. Stock 1932. The Felidae of Rancho La Brea. *Carnegie Inst. Publ.* 422.

Miller, R. L. and J. M. Weller 1952. Significant comparisons in paleontology. *Jour. of Paleont.* 26:6.

Moore, H. B. 1935. A comparison of the biology of *Echinus esculentus* in different habitats. Part II. *J. Mar. Biol. Ass.* New Ser. 20.

Newell, N. D. 1949. Phyletic size increase—an important trend illustrated by fossil invertebrates. *Evolution* 3:2.

Newell, N. D. 1952. Periodicity in invertebrate evolution. In Distribution of evolutionary explosions in geologic time. A symposium. *Jour. of Paleont.* 26:3.

Notini, G. 1948. Hararna. Svenska Djur: Däggdjuren. Norstedt, Stockholm.

Olson, E. C. and R. L. Miller 1951. Relative growth in paleontological studies. *Jour. of Paleont.* 25:2.

Osborn, H. F. 1918. Equidae of the Oligocene, Miocene, and Pliocene of North America, iconographic type revision. *Mem. Amer. Mus. Nat. Hist.,* New Ser. 2:1.

Parr, A. E. 1926. Adaptiogenese und Phylogenese; zur Analyse der Anpassungserscheinungen und ihre Entstehung. *Abh. Theor. Organ. Ent.* 1.

Pearl, R. 1940. Introduction to medical biometry and statistics. Philadelphia: Saunders.

Pearl, R. and A. B. Clawson 1907. Variation and correlation in the crayfish. *Carnegie Inst. Publ.* 64.

Pearl, R. and J. R. Miner 1935. Experimental studies on the duration of life. XIV. The comparative mortality of certain lower organisms. *Quart. Rev. Biol.* 10.

Pearson, Helga S. 1928. Chinese fossil Suidae. *Pal. Sinica C* 5:5.

Pearson, O. P., M. R. Koford and A. K. Pearson 1951. Reproduction of the lump-nosed bat *(Corynorhinus rafinesquei)* in *California Jour. Mamm.* 33:3.

Petersen, B. 1950. The relation between size of mother and number of eggs and young in some spiders and its significance for the evolution of size. *Experientia* 6:3.

Pilgrim, G. E. 1931. *Catalogue of the Pontian Carnivora of Europe in the Department of Geology.* London, British Museum (Nat. Hist.).

Pilgrim, G. E. 1932. The fossil Carnivora of India. *Pal. Indica,* New Ser. 18.

Reinig, W. F. 1938. Elimination und Selektion. Jena.

Reinig, W. F. 1939. Besteht die Bergmannsche Regel zu Recht? *Arch. f. Naturgesch,* N. F. 8.

Rensch, B. 1924. Das Deperetsche Gesetz und die Regel von der Kleinheit der Inselformen als Spezialfall des Bergmannschen Gesetzes und ein Erklärungsversuch derselben: eine Hypothese, *Z. ind. Abst-Vererb.-Lehre,* 35.

Rensch, B. 1929. *Das Prinzip geographischer Rassenkreise und das Problem der Artbildung.* Berlin.

Rensch, B. 1938. Bestehen die Regeln Klimatischer Parallelität bei der Merkmalsausprägung von homoothermen Tieren zu Recht? *Arch. f. Naturgesch.* N. F. 7.

Rensch, B. 1939. Klimatische Auslese von Grössenvarianten. *Arch. f. Naturgesch.* N. F. 8.

Rensch, B. 1947. *Neuere Probleme der Abstammungslehre. Die transspezifische Evolution.* Stuttgart: Ferdinand Enke.

Ringstrom, T. 1924. Nashörner der Hipparion-Fauna Nord-Chinas. *Pal. Sinica C* 1:4.

Robb, R. C. 1936. A study of mutations in evolution. III. The evolution of the Equine foot. *Jour. Genetics* 33.

Rubner, M. 1908. Das Problem der Lebensdauer und seiner Beziehungen zu Wachstum und Ernährung. Munich and Berlin.

Schlaikjer, E. M. 1935. Contributions to the stratigraphy and paleontology of the Goshen Hole area, Wyoming. Part 4: New vertebrates and the stratigraphy of the Oligocene and early Miocene. *Bull. Mus. Comp. Zool. Harvard* 76.

Schlosser, M. 1901. Die fossilen Säugethiere Chinas. *Abh. Bayr. Akad. Wiss. II Cl.* 22:1.

Schlosser, M. 1921. Die Hipparionenfauna von Veles in Mazedonien. *Abh. Bayr. Adad. Wiss.* 29:4.

Sefve, I. 1927. Die Hipparionen Nord-Chinas. *Pal. Sinica C* 1:1.

Simpson, G. G. 1936. Data on the relationships of local and continental mammalian faunas. *Jour. of Paleont.* 10:5.

Simpson, G. G. 1941a. Large Pleistocene felines of North America. *Am. Mus. Nov.* 1136.

Simpson, G. G. 1941b. Range as a zoological character. *Amer. Jour. Sci.* 239.

Simpson, G. G. 1944. Tempo and mode in evolution. *Columbia Biol. Ser.* 15.

Simpson, G. G. 1949. *The Meaning of Evolution.* New Haven: Yale University Press.

Simpson, G. G. 1950a. Evolutionary determinism and the fossil record. *The Scientific Monthly* 71:4.

Simpson, G. G. 1950b. Some principles of historical biology bearing on human origins. *Cold Spring Harbor Symp. Quant. Biol.* 15.

Simpson, G. G. 1951. *Horses.* New York: Oxford Univ. Press.

Simpson, G. G. 1952. Periodicity in vertebrate evolution. In Distribution of evolutionary explosions in geologic time. A symposium. *Jour. of Paleont.* 26:3.

Simpson, G. G. and Anne Roe 1939. *Quantitative Zoology.* New York.

Smith, H. S. 1935. The role of biotic factors in the determination of population densities. *Jour. Econ. Entom.* 28.

Solomon, M. E. 1949. The natural control of animal populations. *Jour. Anim. Ecol.* 18.

Spjeldnaes, N. 1951. Ontogeny of *Beyrichia jonesi* Boll. *Jour. of Paleont.* 25:6.

Spooner, G. M. 1947. The distribution of *Gammarus* species in estuaries. Part I. *Jour. Mar. Biol. Ass.* 27:1.

Stirton, R. A. 1940. Phylogeny of North American Equidae. Univ. Calif. Publ., Bull. Dept. Geol. Sci. 25.

Stirton, R. A. 1947. Observations on evolutionary rates in hypsodonty. *Evolution* 1:1.

Timofeeff-Ressovsky, N. W. 1940. Mutations and geographical variation. In: J. Huxley, ed. *The New Systematics.* Oxford: Oxford Univ. Press.

Twenhofel, W. H. 1950. *Principles of Sedimentation.* London.

Venge, O. 1950. Studies of the maternal influence on the birth weight in rabbits. *Acta Zool.* 31. Stockholm.

Weldon, W. F. R. 1901. A first study of natural selection in *Clausilia laminata. Biometrika* 1.

Wheeler, J. F. G. 1934. On the stock of whales at South Georgia. *Discov. Rep.* 9.

Whiteley, M. A. and K. Pearson 1900. Data for the problem of evolution in man. A first study of the variability and correlation in the hand. *Proc. Roy. Soc.* 65.

Wills, L. J. 1951. *Palaeogeographical Atlas*. London and Glasgow: Blackie.

Wiman, C. 1913. Uber die palaeontologische Bedeutung des Massenster-bens unter den Tieren. *Pal. Zeitschr.* 1. Berlin.

Wright, Sewall, 1932. General, group and special factors. *Genetics* 17.

Wright, Sewall, 1945. Tempo and mode in evolution. A critical review. *Ecology* 26.

Wright, Sewall, 1949. Population structure in evolution. *Proc. Amer. Phil. Soc.* 93:6.

Zdansky, O. 1924. Jungtertiäre Carnivoren Chinas, *Pal. Sinica C* 2:1.

Zdansky, O. 1927. Weitere Bemerkungen über fossile Carnivoren aus China. *Pal. Sinica C* 4:4.

II
ALLOMETRY AND
EVOLUTION

TWO

Observations on Allometry in Mammalian Dentitions; Its Interpretation and Evolutionary Significance

INTRODUCTION

THE GENERAL SHAPE of the mammalian tooth crown appears to be genetically determined in the form of a growth pattern that may or may not include a gradient. If there is no gradient, the shape of the tooth crown will be identical regardless of its absolute size (isometry). This is a special case of allometry *s.l.;* allometry *s.str.,* the presence of a gradient, will always result in differences in shape between teeth of unequal size. The allometry concept was formulated by Huxley (1932).

I have sampled a large body of data, mainly relating to the dentitions of carnivores. In the present paper I intend to discuss some general principles in connection with the study of allometry in mammalian dentitions, relating to biometric technique as well as to interpretation, and illustrated by means of selected examples showing the manner in which such studies may contribute to the theory of evolution.

Reprinted, with permission, from *Acta Zoologica Fennica* (1954) 85:1–13.

THE DESCRIPTION AND INTERPRETATION OF ALLOMETRY

It has been shown (Sinnot 1936) that an allometric pattern is genetically determined and may suffer a change through mutation. In order to obtain a basis for subsequent investigation, we shall initially discuss a case in which the allometry appears to be homogeneous for a population—that is, where biometric method cannot demonstrate any heterogeneity.

Fig. 2.1 represents the covariation between the length and the (talonid) width of the lower carnassial in *Meles meles* (data from Degerbøl 1933: a sample of recent specimens from Denmark and elsewhere, and a sample of subfossil specimens from Denmark). It appears that there existed no difference in this allometry between the subfossil and recent populations, though there is a marked difference in average size (this is one of the rare instances when phyletic growth has occurred in postglacial time: see Degerbøl 1933:635).

Our primary task now is to formulate an accurate description of the relationship between length and width, expressing, in numerical terms, the genotypic correlation between the two variates.

Obviously the phenotypic correlation is not absolute: the variate pairs are not on a single curve. The correlation between the logarithms of the data is $r = .799$. We shall assume, as a working hypothesis, that the deviations are due to phenotypic modification only. We shall further assume that the genotypic correlation conforms to the principle of allometry. Hence the relationship may be expressed by a single curve with the equation

$$y = bx^k,$$

where y and x are the variates, b is the constant of integration, and k the constant of allometry. On a logarithmic grid this curve takes the form of a straight line.

If such a curve is to be fitted accurately, arithmetics must be employed; the scatter is too great to permit accurate fitting by eye. In such cases the method of least squares is commonly used. It gives two regression lines, the regression of y on x, and that of x on y (fig. 2.1); these will diverge, the

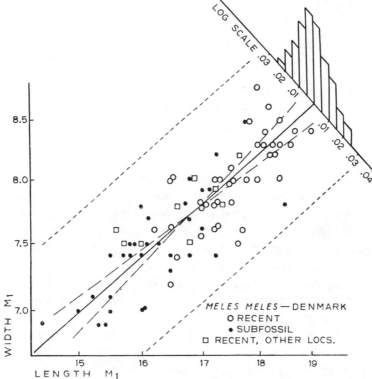

Fig. 2.1. Covariation of (talonid) width and length of M_1 in subfossil and recent badgers, as labelled; data from Degerbøl. Two least squares and one intermediate regression fitted to the data as described in the text. Histogram (top, right) shows distribution around intermediate regression; dashed lines indicate standard range of distribution.

more the lower the correlation. Either one of these regressions, however, is less likely to approximate to the actual genotypic regression than is an intermediate line. The genotypic regression being the subject of interest to us, we must therefore try to find a better approximation.

With absolute correlation the two least squares regressions coincide. This limit regression may, however, also be obtained by the simple formula

$$k = \frac{\sigma_{\log y}}{\sigma_{\log x}}$$

or, when convenient, any development of this formula, such as

$$k = \left[\frac{m_{\log y}^2}{m_{\log x}^2} \right]^{1/2}$$

where m^2 denotes the variance, or second moment around the mean. The resulting curve will be intermediate between the two least squares regressions when the correlation is not absolute; it is given as the main regression in fig. 2.1, where it evidently appears to approximate to the actual genotypic relationship better than the other curves. Hence this method—which has the additional merit of being much less laborious than the method of least squares—has been used throughout. Naturally caution is necessary in its use, since it will give a "regression" even where there is no correlation, and it does not discriminate between positive and negative slope. The results should always be checked by means of scatter diagrams, and, when necessary, by calculating the coefficient of correlation.

Returning now to the case represented in fig. 2.1, we may note that the width of the lower carnassial in *Meles meles* is negatively allometric to its length, but very slightly so ($k = .88$). On an average, larger teeth tend to be relatively more slender than smaller ones, but the difference is almost imperceptible for the present range of size, and is almost overshadowed by the deviations away from the regression.

These deviations were tentatively assumed to be due to modification only. The question may be further studied by investigating the pattern of deviation away from the line.

The frequency distribution of these deviations is summarized in the inserted histogram (fig. 2.1), which shows the distribution to conform excellently to the "normal" binomial type. This appears to support our assumption. If there had been two parallel and distinct regressions involved, our frequency distribution would have been bimodal; if the regressions had not been parallel, it would have been strongly platykurtic. Theoretically, the distribution may still result from intermediate polygenic inheritance comprising a number of distinct regressions close to each other, but there does not appear to be any necessity for this assumption, and it is made definitely improbable by other data.

It appears thus that we are justified in drawing the following inferences:

(1) The subfossil and recent *Meles meles* exhibit the same allometry; (2) the deviations from the regression result from modification; and, hence, (3) there has been no genetic change, as regarding this character, in the *Meles meles* population from postglacial to recent times. The single genetic change to be deduced from the present data is that constituting size increase; this has caused the population to slide up the regression. There is more to be said about the allometry in this tooth, but this will be deferred to a later section.

THE DETECTION OF MUTATION:
SINGLE ABERRANTS

The detection of actual genetic changes in allometry is not always easy. As regarding single aberrant individuals in otherwise homogeneous populations, there is a possibility if the deviation is really great (in which case the difference is often, though by no means always, easily visible to the trained eye, even without biometric analysis). But judgment from morphological inspection only may be fallacious: aberrations may result also from strong allometry, if especially large or especially small variants happen to turn up. From the frequency distribution around the regression line we may be able to form a tentative judgment as to the significance of a deviation. Simpson (1941) considers the estimate of variation limits in a sample of 1,000 (which is $= M \pm 3.24\sigma$) to give reasonable security; this is called the standard range. In the present case, the standard deviation of the frequency distribution around the regression is $\sigma = .01015$, in terms of logarithms. The standard range would, then, be indicated by lines parallel to the regression and at a distance of .033 (logs) from it (fig. 2.1). Observations at a greater distance from the regression may be considered to indicate genetic change. Actually, this is valid only in the vicinity of the means for the two variates, x and y; the standard error of the regression coefficient is not taken into account. This may, however, be allowed for if the necessity arises; but in doubtful cases it is better to try and find other allometry relations and see how they behave.

POLYMORPHISM

The standard deviation of the present distribution, in logs, was = .01015. A number of such standard deviations have been studied in order to form an estimate of the average magnitude of the log deviation for genetically homogeneous populations. It was found that the values, for functional teeth, were fairly constant and varied between .010 and .015 (these data relate to mensuration in different planes, such as length and width; much higher correlations, i.e. lower sigmas, may be found in, for instance, the covariation of two length dimensions). For vestigial teeth, where the shape is less rigidly determined, the deviations were often much greater; this can hardly be due to genetic diversity in general, for it would be peculiar if all vestigial teeth were controlled by several allometry genes, when most functional ones, apparently, are not.

In a few instances, however, functional teeth showed deviations greatly exceeding the usual values. Some instances are being published by me (1954) and will not here be discussed in detail. It was found that the homologous tooth, in other populations, exhibited "normal" scatter, and it was inferred that the great increase in variation resulted from the presence of different allometry alleles. In some instances the alleles could be tentatively homologized respectively, with genes present in other populations. The best instances (see article 3) were found in the first upper molar of recent and fossil bears, where two types of allometry are in evidence, in different proportions, in various populations.

INTERGROUP DIFFERENCE

Clear evidence of genetic change is found when the allometry is decidedly different in related populations. This appears to be a very common state of matter in closely related but distinct species. One instance is given in fig. 2.2; this is the covariation of paraconid height and crown length in the lower carnassial of small cats, after data from Degerbøl (1933).

In the wildcat (*Felis silvestris*, a series of subfossil specimens from Denmark) paraconid height is positively allometric to crown length (k = 1.39). The distribution around the regression line (inserted histogram) does not indicate heterogeneity; σ = .015. In the domestic cat (*Felis catus*), the

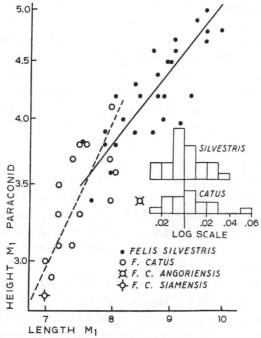

Fig. 2.2. Covariation of paraconid height and length of M_1 in subfossil *Felis silvestris* and recent *F. catus*, as labelled; data from Degerbøl. Inserted histograms show distributions around regression for each sample.

allometry is also positive and much stronger (k = 2.17, one of the highest values I have found in mammalian dentitions). The Siamese appears to be on the same regression; *angoriensis* is probably not, and was left out in calculation. The distribution, then, has the parameter σ = .014. We may infer that the domestic cat and the European wildcat differ genetically in this respect; and that the wildcat population is "pure" for one allele and the domestic form for another, with the tentative exclusion of *angoriensis*.

OBSERVATIONS ON HYPSODONTY

In this case, as, for instance, with the bear M_2, to be discussed below, height is positively allometric to length, and therefore larger teeth tend to be more hypsodont. It is, perhaps, superfluous to add that this relation is not universal for mammalian tooth crowns; all sorts of allometry have been found, and the trends may be of bewildering variety in closely related species (Kurtén 1954). It may, however, happen that there is a direct intrapopulation correlation between hypsodonty and gross size. As this is contrary to earlier statements by Simpson (1944) and myself (1953), some further remarks are necessary. On the basis of studies on horse teeth, Simpson (1944:p7–8) stated: "Hypsodonty, the relationship between vertical and horizontal dimensions, is positively correlated with size and with most linear dimensions among successive populations, but shows no such correlations among individuals or among contemporaneous populations. . . . The stated independence is an evident and, I believe, incontrovertible biological fact." In 1953 I followed Simpson without entering on analysis of this special question. From the data in the previous paragraph (and also in the next paragraph; and from a number of unpublished data sampled by me) it is, however, clear, that the notion of independence as a general principle is erroneous.

This does not reflect on Simpson's main thesis, which is that genetic changes have occurred in horse history affecting the length-height relation of the cheek teeth. The manner in which these changes were realized was probably analogous to the instances illustrated in the present paper; as the raw data are not given, detailed allometry analysis is not possible.

ALLOMETRY AND SELECTION

What is the biological function of allometry, as opposed to isometry, in a tooth crown? As far as I can judge, almost none at all. Of course the greater hypsodonty of a large individual, when crown height is positively allometric, may be useful, under certain circumstances; but in that case the corresponding brachyodonty of a small variant would seem to be harmful.[2] I do not think that the allometric trend *per se* has any biological function (except, perhaps, a preadaptive one in exceptional instances); what is important is, I think, the *average relation* between two dimensions (say, length and width) that is conditioned by a certain regression and a certain average size. What is selected for appears to be a certain average proportion, regardless of the resulting shape and proportions of distal variants, except perhaps where these latter become biologically *untragbar* as a result of excessive growth gradients. The average proportion is realized by means of the available genetic material, and this, in turn, may be isometric or allometric.

This is especially clear in cases where related populations show differences in allometry. The necessity for a change in allometry with change in average size is easily appreciated if we imagine the allometry of one population projected into the gross size of the other. Thus, for instance, it is evident that the pattern of *Felis catus* (fig.2.2) cannot persist unchanged with phyletic growth or phyletic dwarfing. In the former case, the trend would lead to an enormously hypsodont carnassial; in the latter case, the carnassial would grow very low. Both changes would apparently be inadaptive. Accordingly, the larger species, *Felis silvestris,* is on a different regression, and is much less hypsodont that it would have been with the allometry of *catus.*

Fig.2.3 shows another case, the covariation between protoconid height and crown length of M_2 in *Ursus spelaeus* (original data: the von Nordmann collection from Odessa, at the Geological Institute of Helsingfors University) and *U. arctos* (original data: recent Finnish population, collections at the Zoological and Anatomical Institutes of Helsingfors University). In both species there is a positive allometry of height on length; yet the average relation between the two dimensions is identical, as will appear from the mean indices (table 2.1).

This relation is apparently optimal for the tooth crown in question, but its realization has necessitated a genetic change somewhere in the ancestry of either species, or both. The actual history of this change may be mapped in some detail and is essentially similar to that of M^1, where stability in average proportions is similarly brought about by the interplay of two different regressions (Kurtén, unpublished). Suffice it for the moment to note the important fact that *similarity,* in the present case, *has sprung from genetic change, where stability* (except for size) *would have produced dissimilarity.* And this, again, delivers a *coup de grace* to *all studies trying to evaluate relationships by means of indices or ratios* without previous analysis of allometry.

Instances such as this may be called micro-evolutionary examples of convergence, in the almost literal sense of the word—convergence in ratios, that is, shape. Studies utilizing relative values—indices—will, in all such instances, give a completely misleading picture of what actually has hap-

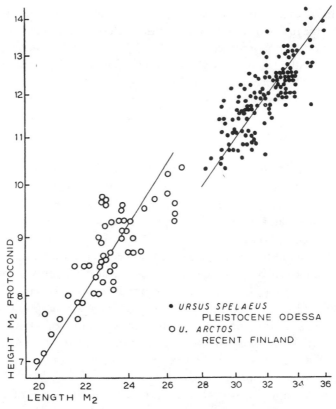

Fig. 2.3. Covariation of protoconid height and length of M_2 in *Ursus spelaeus* (Pleistocene, Odessa) and *U. arctos* (recent, Finland); original data.

pened, and of the true genetic affinity of populations; whereas the study of allometry means a new and important tool for the detailed mapping of microevolutionary events, and for the evaluation of actual genetic relationships of extinct forms where no experiments can be performed.

TEMPORAL SUBSPECIFIC DIFFERENTIATION

Let us finally consider a case in which a subspecific advance during a geologically short period has been detected by more orthodox methods. We return to the M_1 of *Meles meles*, the length-width relation of which was

Table 2.1.
Comparisons of Height and Length in *U. arctos* and *U. spelaeus*

	mean		index
	length	height	100 height/length
U. arctos	23.2	8.3	37
U. spelaeus	31.9	11.9	37

illustrated in fig. 2.1. Degerbøl (1933), when making a subspecific distinction between the oldest subfossil badgers and the recent Danish ones, pointed not only to the difference in size, but also to the difference in relative size between trigonid (the anterior shearing part) and talonid (the posterior crushing part). The trigonid is relatively smaller, and the talonid relatively larger, in the recent form, which is thence more specialized for the particular mode of life of this species. Does this difference result from simple allometry, or has there been a genetic change?

The answer appears to be perfectly clear and definite: there has been a genetic change, and the subspecific distinction is justified beyond dispute. Fig. 2.4 shows the covariation between trigonid length (measured inter-

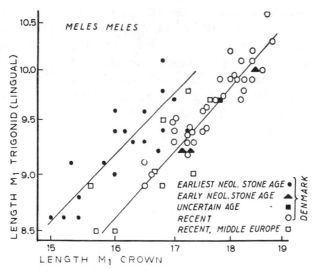

Fig. 2.4. Covariation of trigonid length (lingual, to metaconid) and crown length of M_1 in subfossil and recent badgers, as labelled; data from Degerbøl.

nally, from anterior end to metaconid) and crown length. It shows two distinct regressions, one for the oldest subfossil sample with persistently longer trigonid, and another for the others, with shorter trigonid.[3] In spite of the difficulties of accurate mensuration, pointed out by Degerbøl, the scatter is not very great. It may be noted that the actual change occurred between the periods labeled "earliest" and "early neolithic stone age"— that is, the population was completely saturated by the new gene in a very short time, and has been stable since then. The recent Danish population has evidently lost the "long trigonid" gene completely, whereas it may persist in some Middle European populations. The rapidity of the change may have been made possible by the fact that the Zealand population evidently was isolated during the period in question (Degerbøl, 1933) and thus would have permitted a relatively short *Anlaufzeit* of the new gene.

CONCLUSIONS

It has been shown that the study of allometry permits the actual mapping of microevolutionary events in geologic time. Thus we may, for good, reject the claim repeatedly stated by neozoologists (e.g., Huxley 1942), that data of a paleontological nature cannot give evidence on the mechanism of evolution. Data of this sort shed light on selection and indicate the essentially "opportunistic" nature of evolution (Simpson 1949). They are throughout definitely in favor of Neodarwinism, and flatly incompatible with "orthogenetic" finalist, and Lamarckist theories of evolution. *No other interpretation than that of random mutation and selection is consonant with these data, which are essentially paleontological in nature.*

NOTES

1. There is, however, a possibility. Changes in allometry may be brought about by pleiotropic genes that are selected for in conjunction with other qualities. Balanced polymorphism for some other quality may, thus, be reflected in polymorphism for allometry; the latter may not be of sufficient functional disadvantage, when appearing in a vestigial tooth, to bring about adverse selection, or to be eliminated by modifiers.

2. There is evidently more to be said on this topic. I have, however, scant data for herbivores, and shall not enter into detail.

3. The difference appears also from other data published by degerbøl, though not in all possible combinations.

REFERENCES

Degerbøl, M. 1933. Danmarks pattedyr i fortiden i sammenligning med recente former. I. *Vid. Med. Dansk Naturh. For.* 96, Festskr. II.

Huxley, J. S. 1932. *Problems of Relative Growth*. London.

Huxley, J. S. 1942. *Evolution: The Modern Synthesis*. London.

Kurtén, B. 1953. On the variation and population dynamics of fossil and recent mammal populations. *Acta Zool. Fennica* 76.

Kurtén, B. 1954. The type collection of *Ictitherium robustum* (Gervais, ex Nordmann) and the radiation of the ictitheres. *Acta Zool. Fennica* 86.

Simpson, G. G. 1941. Range as a zoological character. *Amer. Jour. Sci.* 239.

Simpson, G. G. 1944. *Tempo and Mode in Evolution*. New York: Columbia University Press.

Simpson, G. G. 1949. *The Meaning of Evolution*. New Haven: Yale University Press.

Sinnot, E. W. 1936. A developmental analysis of inherited shape differences in cucurbit fruits. *Amer. Nat.* 70.

THREE

Some Quantitative Approaches to Dental Microevolution

TOOTH SIZE AND BODY SIZE

T HE LONG DIAMETER of a tooth seems often to be a good measure of size in general, for if care is taken to avoid dimensions affected by wear or continuous growth, this measure will be constant regardless of individual age.

Plotting the size of teeth against time is often the best way to study evolutionary changes of size in mammal populations (fig.3.1). It has been said that the apparent dwarfing of many mammalian species in postglacial times is spurious and really due to limited longevity in the living population, where animals die before reaching maximal size. The comparison of dental measurements, however, proves that many species have in fact been dwarfed in the late Quaternary (Kurtén 1959).

This presupposes that there is really a correlation between the size of teeth and the size of the animal as a whole. Such is true within populations, as exemplified by the positive correlation between dental size and skull or jaw length in four examples taken at random from a large collection of measurements (table 3.1). Of course the correlation is not absolute; comparatively small individuals may have relatively large teeth, and vice versa. The former type of deviation was more frequent and is due partly to the inclusion of animal specimens that had not yet grown to maximal size, though care was taken to exclude all obviously immature specimens.

Reprinted, with permission, from *Journal of Dental Research* (1967) 46:817–828.

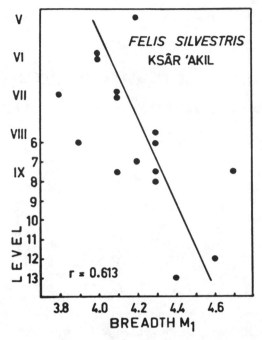

Fig. 3.1. Relationship between stratigraphic level and width of lower carnassial in wild cat (*Felis silvestris*) from the rock shelter of Ksar Áikl, Lebanon. The sequence covers the entire last glaciation (Würm). The correlation is highly significant, showing that a decrease in size took place during deposition. (Kurtén, 1965).

Table 3.1.

Correlation Between the Length of a Cheek Tooth and That of the Skull or Jaw in Local Populations of Four Species of Carnivores

Species and Locality	No. of Specimens	r	P
Vulpes vulpes, Denmark (P⁴ skull)	43	0.58	<0.001
Felis silvestris, Scotland (P⁴, skull)	48	0.62	<0.001
Crocuta crocuta, Balbal, Tanganyika (P⁴, skull)	70	0.47	<0.001
Ursus spelaeus, Mixnitz, Austria (M², mandible)	42	0.63	<0.001

Data on *V. Vulpes* from Degerbol (1933), other data is original, obtained mostly from collections in the British Museum (Natural History) and the Paleontology Department of the University of Vienna.

ALLOMETRY BETWEEN TEETH

The relationship between body and tooth size may undergo various changes as the absolute size increases or decreases. For instance, among carnivores, large forms tend to have relatively larger canine teeth than related small forms have (fig.3.2). Within populations, different teeth usually tend to have an allometric relationship to each other. If their size is plotted on a logarithmic graph, the values will cluster around a straight line that may be isometric (fig.3.3) but seems often to be more or less allometric (fig.3.4). In the former instance, the relationship between two teeth will be stable whatever the absolute size. In the latter (and seemingly more common) instance, the size relationship between the teeth will change in a regular way with gross body size. For example, the third lower molar of the bear tends to be positively allometric to other molars and thus become relatively longer in larger individuals.

The possible adaptive significance of such allometries is obvious. With change in absolute size, the lever arms of the jaws change in direct relation to the linear measurements, but the masticating surface of the teeth and the cross section (and power) of the muscles change in relation to their squared values, and the bulk of the ingested food, and of the body to be fed, change in relation to their cubed values. Adaptation thus requires

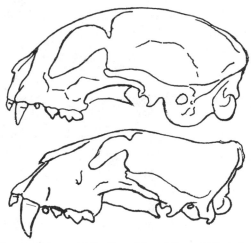

Fig. 3.2. Skulls of wild cat *(Felis silvestris)* and lion *(Felis leo)*, brought to same length, to illustrate difference in relative size of canine teeth.

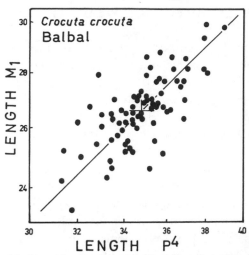

Fig. 3.3. Double logarithmic plot of the lengths of upper and lower carnassials in a local population of spotted hyena (*Crocuta crocuta*) from Balbal, Tanganyika. The coefficient of allometry is $k = 1.01$, not significantly different from unity or isometry.

complex changes in the relationships between different parts of the dentition, whenever great changes in average size occur. It would be an advantage to have the appropriate allometries built into the genotype, especially in such forms that show extensive evolutionary plasticity in size. The bears illustrate this beautifully, as they are extremely variable in size (Kurtén 1955, 1959, 1965).

CORRELATION BETWEEN TEETH

The allometric relationships between teeth imply that they show positive correlation in regard to size. There are, of course, many phyletic instances in which it would seem that some teeth increase in size at the cost of the teeth next to them. Rensch (1954) mentions several examples, among which the sabertoothed cats may be the best known (fig.3.5). In usual cats, the large size of the canine tooth contrasts with the small premolar (P^2) just behind it. In the sabertoothed tiger with its enormous sabers, P^2 had vanished entirely and P^3 was much reduced. There is, then, a negative phyletic correlation between the size of the canine and that of the anterior premolars in the history of the sabertoothed tigers.

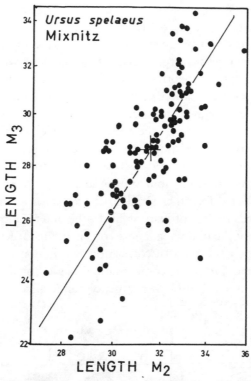

Fig. 3.4. Double logarithmic plot of the lengths of second and third lower molars in a local population of the cave bear *(Ursus spelaeus)* from the late Pleistocene of the Dráchenhöhle at Mixnitz in Austria. The coefficient of allometry is $k = 1.62$, M_3 being positively allometric to M_2.

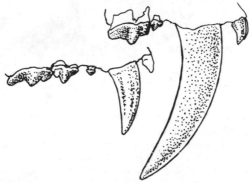

Fig. 3.5. Upper tooth row in the lion *(Felis leo)* and a saber-toothed tiger *(Smilodon fatalis)* show loss of anterior premolars in conjunction with the growth of the canine tooth, interpreted as compensation.

Within any single population, however, this correlation is not negative. Generally speaking, teeth are positively correlated in size, and this correlation tends to be greater the closer the teeth to each other, to become maximal between teeth that actually occlude with each other (fig.3.6).[1] The adaptive importance of this is self-evident. Teeth have to maintain a reasonably good fit in occlusion to work properly, and consequently must retain a constant size relationship to each other. Thus there arises a field effect: the teeth are in a correlation field, the strength of which is proportional to distance. In some instances, the field can be shown to have turned back on itself so that end members in a series show increased correlation. Similar fields with the same properties have been established for other serial structures, such as the phalangeal series of the hands or feet of man or birds, the antennae of insects, the primary feathers of the bird wing, and so on (Bader 1960; Kurtén 1953).

It is evident that the positive correlation will inhibit changes of the compensation type. However, compensation may occur if the allometry axis becomes transposed through a genetic change affecting the growth

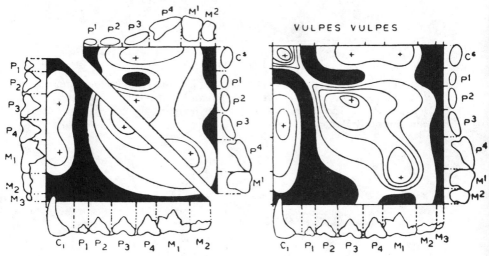

Fig. 3.6. Correlation fields in dentition of the rex fox *(Vulpes vulpes)*. Left: Correlations within mandibular and maxillary tooth rows, respectively. Right: Correlations between maxillary and mandibular teeth. The strength of the correlation, expressed according to Fisher's z test, is represented by contour lines with interval 0.10; black fields indicate $z < 0.40$ and not significant. Effect of sex dimorphism has been eliminated. (Kurtén 1953).

field. A typical instance (fig.3.7) is the covariation between the lengths of the second and third lower premolars in two temporal populations of the Pleistocene cave hyena *(Crocuta crocuta)* from England. In the Eemian or last interglacial form, which lived more than 60,000 years ago, P^2 was comparatively large in relation to P_3. A definite shift in the allometry axis occurred in the evolution of the Wurmian (last glaciation) form, which existed about 60,000 to 10,000 years ago. The second premolar became shorter; the third became longer. This appears to be a true compensation effect: What P_2 lost in the transposition was gained by P_3. Perhaps the more spectacular instances of compensation may be resolved into sequences of such transpositional steps, in which one tooth increases at the cost of its neighbors.

It should also be noted that vestigial teeth show little or no correlation with the field of the other teeth (e.g., fig.3.6, M^2 and especially M_3) and so are free to be reduced and lost independently, which is also an evident advantage from the point of view of adaptation.

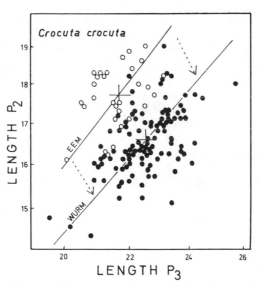

Fig. 3.7. Double logarithmic plot of the lengths of second and third lower premolars in two temporal populations of the cave hyena *(Crocuta crocuta)* in England (southern Devonshire). White circles = Eemian, Hyena Stratum, Tornewton Cave; black circles = Würmian, Cave Earth, Kent's Cavern; arrows = transposition of allometry axis.

INTRADENTAL ALLOMETRY

The shape of a tooth is also governed by allometry. The height and length relationships of the lower carnassial in cats is shown (fig.3.8). The height gives strong positive allometry in the domestic cat, so that a large M_1 will have a much higher crown than a small M_1. Positive allometry of height and width over length appears to be almost the rule in most cheek teeth, and this is natural for physiologic reasons. Increase of the height and width increases the mass of the crown and its durability when worn, corresponding to the increase in bulk of the animal relative to the masticatory surface of the teeth.

It is interesting that the allometry of the domestic cat is offset in relation to that of the wild cat. It is natural that a strong allometry of

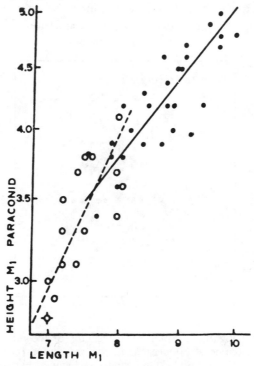

Fig. 3.8. Double logarithmic plot of anterior paraconid height and crown length of lower carnassials in domestic cats (white circles) and subfossil remains of wild cat *(Felis silvestris)* from Denmark (black circles); the Siamese cat is also shown (circled cross). The height is positively allometric: $k = 2.17$ for the domestic cat, $k = 1.39$ for the wild cat. (Kurtén, 1954.)

this type cannot continue unchanged for a great range of size, because this would result in a monstrosity. If the average size of the animal changes much, a change in the allometry is necessary. This may sometimes happen, as in the cat, by a change of the allometric slope. A transposition without great change of slope, however, is apparently more frequent (fig.3.9). The height-length allometry of the second lower molar in the cave bear is transposed in relation to that in the brown bear.

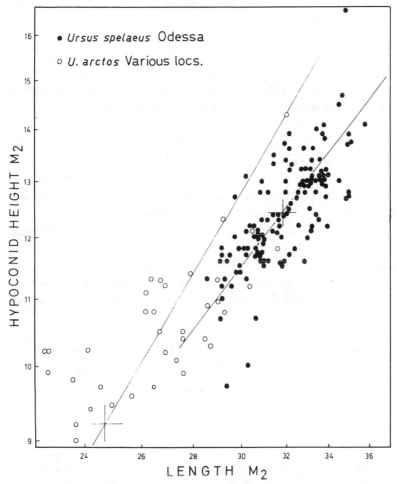

Fig. 3.9. Double logarithmic plot of hypoconid height and crown length of second lower molar in recent brown bears, *Ursus arctos* (white circles), and late Pleistocene cave bears, *Ursus spelaeus* (black circles) from Odessa. One of the latter is monstrously high crowned (see fig. 3.10) and corresponds to the growth pattern in *U. arctos*.

One specimen from a cave bear is clearly off the main regression line, and instead lies on the extension of the brown bear line; this Odessa specimen appears to have been formed in accordance with the growth pattern of *Usus arctos* and is, therefore, a morphologic monstrosity (fig.3.10). One similar instance was found in the third molar in the same Odessa sample (fig.3.10). Both specimens are isolated teeth of 1-year-old bear cubs and may have come from a single cub. It has been suggested that these represent mutants to an ancestral type that survived unmodified in *U. arctos* (Kurtén 1955).

The reverse condition, *spelaeus*-like forms in brown bear, has also been observed (fig.3.11). Three recent specimens from brown bear (two from Finland, one from the Pyrenees) departed widely from the norm in *U. arctos* and clustered around the *Ursus spelaeus* trend line.[2] Earlier populations show various distributions of intermediate type (Kurtén 1955).

TALONID LENGTHS IN THE BADGER

In the simplest form, the mechanism of a transposition is apparently a single mutation. This is suggested for the instance of the badger. In the Zealand (Denmark) population, the relative length of the carnassial talonid was suddenly increased during a short interval, probably no more than 1,000 years, in the early Neolithic (fig.3.12). At that time, Zealand was already an island, effectively isolated from the continent and from Sweden. Although the Belts and Öre Sound may freeze over in exceptionally cold

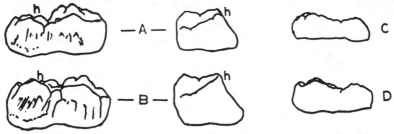

Fig. 3.10. Normal and aberrant lower molars from cave bears *(Ursus spelaeus)*, of Odessa. External and posterior views show a normal M_2 (A) and a M_2 with hyperthrophied hypoconid (B), the aberrant specimen in Figure 9, both the same length; external views show a normal M_3 (C) and an abnormally hypsodont M_3 (D), both the same length. *h* = hypoconid summit. (Kurtén 1955.)

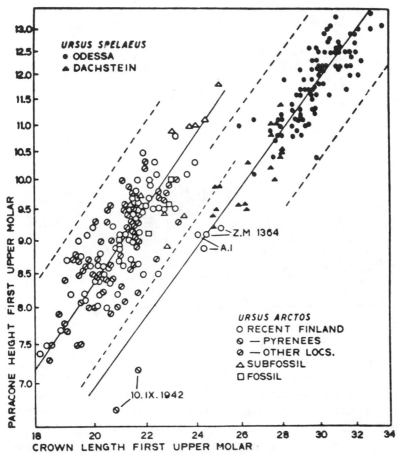

Fig. 3.11.   Double logarithmic plot of paracone height and crown length of first upper molars in recent brown bears, *Ursus arctos* (white circles), and late Pleistocene cave bears, *Ursus spelaeus* (black circles), of Odessa. Three brown bear teeth have extremely low crowns and correspond to the growth pattern of *U. spelaeus*. (Kurtén, 1955.)

winters, the badger is not abroad in severe cold and is not likely to have trekked across the ice. The local population was thus isolated, which would have permitted a comparatively short *Anlaufzeit* for the "long talonid" gene.

At an earlier stage of the postglacial period, Denmark was still connected with Sweden by what is now Öre Sound (Nilsson 1953, 1960). This was

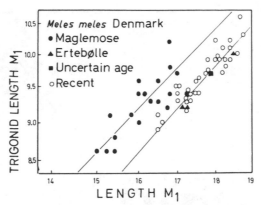

Fig. 3.12.   Double logarithmic plot of trigonid length (measured lingually) and total crown length of lower carnassials in subfossil and recent badgers *(Meles meles)* from Denmark. Notice the transposition from the Maglemosian type with a relatively long trigonid (and hence a short talonid) to the later type with a shorter trigonid (and a long talonid). Black circles = Maglemose period (earliest Neolithic); triangles = Brabrand and Ertebølle periods (early Neolithic); squares = uncertain age; white circles = recent. (Kurtén, 1954.)

the time when the badger immigrated into Zealand and farther north into Sweden and Norway. In Sweden and Finland, the species ranges northward approximately to the 64th parallel; hence, there is no contact between the two populations around the Bothnian Gulf. The badgers of Sweden and Norway thus also form an isolated population. This isolated population, being descended from the early Danish form with a short talonid, might be expected to have conserved that character, and so it has (fig.3.13). The subfossil and recent Swedish forms shown in the scattergram give values well off the "long talonid" axis and cluster around the "short talonid" axis.

The Finnish badger population is connected with the badger population of continental Europe. Here it would seem that both types are present, for the values tend to range along both the short and long talonid axes (fig.3.13).

CARNASSIAL CUSP IN THE RED FOX

A somewhat different type of situation occurs with the red fox *(Vulpes vulpes)*. The history of this species in Denmark was studied by Degerbøl (1933), who reports that the lower carnassial in recent Danish specimens

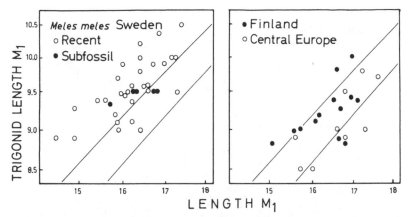

Fig. 3.13. Double logarithmic plots of trigonid length and total crown length of lower carnassials in badgers *(Meles meles)*. Covariation axes are as in fig. 3.12. Left: Material from Sweden (black circles = subfossil material from Ageröd, Scania; white circles = recent material). Right: Recent specimens from Finland (black circles) and central Europe (white circles). (Degerbøl 1933.)

tended to be much broader than in subfossil, Neolithic specimens. When analyzed for width-length allometry, the change takes the form of a simple transposition (fig.3.14). Degerbøl also found that the change in relative width was associated with the development of a small cusp at the internal junction of the trigonid and talonid (fig.3.15). The cusp is absent in the subfossil material and is present in the recent specimens. The two morphotypes may be termed "cuspless" and "cusped."

The distribution of this characteristic was studied in a number of populations and is summarized (table 3.2). Generally, the cuspless morphotype seems to be rare in most populations. Of the recent red fox, no cuspless specimens at all were found in central and western Europe; the trait appears in northern and eastern populations with increasing frequency. During the Würm glaciation, however, the cuspless morphotype is also found to have occurred in central and western Europe. There has been no trace of it in the Eemian interglacial period, but the Saale glaciation population in western Europe has been found to have once more had cuspless M_1 teeth in small numbers. There would seem to be some connection between the presence of cuspless morphotype and a cold or perhaps a continental climate, perhaps due to pleiotropic gene effects.

The entirely cuspless Neolithic population in Denmark perhaps may be regarded as a Pleistocene relict that was later superseded by immigrating

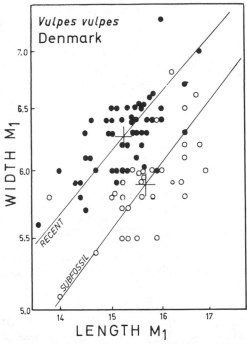

Fig. 3.14. Double logarithmic plot of length and width of lower carnassials in the red fox *(Vulpes vulpes)* in Denmark. Transposition is from a long, narrow type to a shorter, broader type. White circles = subfossil cuspless teeth; black circles = recent cusped teeth. (Degerbøl 1933.)

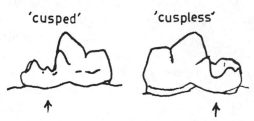

Fig. 3.15 Internal views of lower carnassials in the red fox *(Vulpes vulpes)*. Left: Eemian period, Hyena Stratum of Tornewton Cave (England); left M_1 with well-developed cusp at innter junction of talonid and trigonid. Right: recent, Finland; right M_1 (semiocclusal view) without the cusp.

Table 3.2
Frequency of Cuspless Morphotype in Samples of Lower Carnassial
Molars of Red Fox, *Vulpes vulpes* (Original Data)

Sample	No. of Specimens	Cuspless (%)
Recent		
Great Britain	38	0
Germany	47	0
Sweden	77	11.7 ± 3.7
Finland	102	26.9 ± 4.3
Poland	8	37.5 ± 17.1
Fossil		
Würm, Great Britain	41	7.3 ± 4.1
Würm, Continent	26	15.4 ± 7.1
Würm, Palestine	40	5.0 ± 3.4
Eem, Great Britain	19	0
Saale, Great Britain	73	4.1 ± 2.3

cusped fox. The fox, unlike the badger, will not hesitate to cross an expanse of ice, and the occasional freezing over of the Belts and Ore Sound would allow introduction of cusped fox from Sweden or the continent.

Allometry trend axes were calculated separately for cusped and cuspless M_1 teeth in Finnish and Swedish foxes for comparison with the situation in Denmark (table 3.3). The distribution of the two forms in the Finnish sample is shown (fig.3.16).

The axes for the cuspless teeth are approximately the same in all the populations; comparison by means of the F test fails to reveal any significant difference between the three samples. The patterns found in the cusped teeth are not the same, however; all the three groups differ significantly from each other.

Another feature is especially clear in the Finnish sample: The "spread" of the cusped form around the regression line is much greater than that of the cuspless. This appears in the correlation coefficient, which is lower in the cusped than in the "cuspless" form, and in the standard deviation around the regression line (S_{yx}), which is greater in the cusped than in the cuspless form. The same pattern emerges when the two Swedish samples are compared with each other (table 3.3).

In the Danish red fox population, however, there is no such difference. Although the correlation coefficient is slightly lower in the cusped form, the standard deviation is also lower; this would suggest a more restricted

Table 3.3
Allometry Axes for Cusped and Cuspless M₁ in Samples from Red Fox, *Vulpes vulpes*

Morphotype	No. of Specimens	Mean Log Width	Mean Log Length	k	r	s_{yx}
Denmark						
Subfossil cuspless	33	0.7700 ± 0.0046	1.1951 ± 0.0035	1.31	0.77	0.0174
Recent cusped	49	0.7971 ± 0.0031	1.1826 ± 0.0026	1.17	0.71	0.0156
Finland						
Cuspless	28	0.7888 ± 0.0052	1.2120 ± 0.0047	1.12	0.81	0.0165
Cusped	76	0.7962 ± 0.0028	1.2111 ± 0.0023	1.21	0.50	0.0211
Sweden						
Cuspless	9	0.8123 ± 0.0028	1.2189 ± 0.0019	1.43	0.90	0.0128
Cusped	68	0.8057 ± 0.0023	1.2116 ± 0.0021	1.07	0.72	0.0161

k = Coefficient of allometry; r = coefficient of correlation; s_{yx} = standard deviation around width-length regression line.

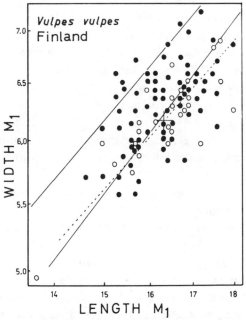

Fig. 3.16.   Double logarithmic plot of length and width of lower carnassials in the recent red fox *(Vulpes vulpes)* of Finland. Axes are continuous as in fig. 3.14. Dotted line = axis for cuspless teeth of the recent red fox of Finland; white circles = cuspless teeth; black circles = cusped teeth.

variation around the regression line. (This apparent contradiction is because k is lower in the cusped form, thus reducing the spread of S_{yx} as expressed in terms of y (log width) as well as slightly reducing the r value.)

The analysis gives the impression that the cusped Finnish and Swedish samples are contaminated, as compared with the cusped Danish sample and all the cuspless samples. The following simple genetic explanation might account for the situation as found, assuming that the genes for cusped and cuspless forms are allelic and have the following phenotypic expression: Homozygous cusped teeth are broad and the cusp is present, as in the entire recent Danish red fox population. Heterozygous cusped and cuspless teeth are narrow and the cusp present, as in majority of Finnish and Swedish cusped M_1. Homozygous cuspless teeth are narrow and the cusp is absent, as in all the cuspless samples.

A more definitive analysis must await the study of pedigrees. Whatever the ultimate solution, it seems evident that the genesis of a tooth cusp is here interconnected in a complex way with a shift in the length-width allometry of the same tooth, and perhaps also with some feature in the environmental tolerance of the organism. In a broader perspective, this might examplify the appearance of "preadaptive" dental characteristics as byproducts of genetic change that is primarily related to some quite different feature.

DARWINIAN SELECTION OF DENTAL CHARACTERISTICS

The effect of differential mortality on the dentition is open to direct study if large enough samples of teeth from a single population are available. Again, dimensions studied should be measured so as not to be affected by wear or growth. Various measures of individual age are available, from the almost uncannily precise age determination produced by physical anthropologists to the comparative wear of the teeth themselves (Kurtén 1953).

When old individuals are compared with young, it can usually be shown that the variation is slightly lower in the old: there is a slight reduction in the value of the standard deviation. This evidently results from greater mortality among distal variants than among proximal. This appears to be so in the cave bear, of which various large local samples have been checked in this way. A comparatively large sample is needed for such studies.

Occasionally, the mean value for the old differs greatly from that in the young, even though effects of wear or growth are excluded. This means that selection has a directional component and favors an optimum different from the population mode. An example of this has been published in detail elsewhere (Kurtén 1957a, 1957b, 1964), and it is necessary only to recapitulate the outline. It concerns the interaction between the paracone of M^2 and its receptacle in occlusion (the notch between the protoconid and hypoconid in M_2) in the cave bear (*U. spelaeus*) of Odessa, from the late Pleistocene. There appears to have been a strong differential mortality directed against all animals with a comparatively long paracone in M^2 or a short distance between protoconid and hypoconid in M_2, or both. (Most of the material consists of loose teeth, which were aged on the basis of root formation and wear.) Thus, the Darwinian selection would favor a situation where the protocone is shorter and the receiving notch is longer than the modal, indicating that there tended to be a slight malocclusion in many cave bears of this local population.

There is not enough data on other populations to make a comparison, but it is not likely, a priori, that the same situation would be found in other demes. The cave bear apparently was a highly stationary form with little interchange between adjacent populations. The Odessa population is geographically marginal.

It has been suggested that the situation might reflect lack of genetic variation in the trait considered, for it seems the selection took no effect. (The material must represent some 10,000 or 20,000 years of continuous settlement by the cave bear.) However, this suggestion has won little popularity among geneticists, who usually invoke balanced polymorphism.

When selection on the tooth cusps is discussed, it is often stated that the cusps are worn smooth in a short time and so details in their shape cannot have any great selective effect. This reasoning overlooks the important fact that the years in which the cusps are first worn off are also the years of peak mortality. In many wild mammal populations, as many as 70 percent die before sexual maturity (Kurtén 1953); subsequently, the mortality is much reduced and increases again as old age begins. Thus, there are two major episodes of Darwinian selection in the life of a cohort: (1) Juvenile mortality, perhaps mainly affecting details of cusp configuration and occlusion, and (2) senile mortality, perhaps mainly affecting the long-term durability of the dentition. Senility, for most mammals in nature, simply means their teeth are worn out, although they may be quite fit in other respects (Flower 1931); hence, characteristics increasing the longevity of the dentition (e.g., increased hypsodonty, thickness and structure of enamel) are at a selective premium in this phase.

Mortality in middle life also may be selective, but it may be postulated that it is usually less directly connected with the dentition than the juvenile and senile mortality.

Darwinian selection may also affect the outline shape of a tooth, as in the third lower molar of the cave bear of Odessa. In this population, those that died young had a definitely shorter and broader M_3 than those that survived to advanced age, so that the width-length regression was gradually displaced in older age groups (fig.3.17; table 3.4). There is no possible wear or growth factor that could result in such a shift. The only change brought by age is the development of an interstitial wear facet against M_2, which may slightly reduce the length in the older specimens; this, however, is antithetic to the selective trend suggested.

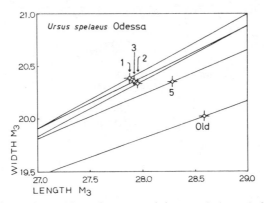

Fig. 3.17.   Change in position of means and the covariation axis for the width and length of M_3 with increasing age (cohort) in the cave bear *(Ursus spelaeus)* from Odessa. Approximate age (years) is indicated by numbers. Selection would appear to favor a longer and narrower type than that modal in the population, because only these survive to form the "old" group (see table 3.4).

Table 3.4

Changes in Covariation of Length and Width of M$_3$ with Increasing Individual Age in a Cohort of the Cave Bear, Ursus spelaeus, from Odessa

Age (Years)	No. of Specimens[a]	Width (mm.)		Length (mm.)		r	b[b]
		Mean	SD	Mean	SD		
I	122	20.39 ± 0.11	1.24	27.89 ± 0.20	2.25	0.67	0.55
2	76	20.34 ± 0.13	1.17	27.96 ± 0.26	2.27	0.75	0.52
3	49	20.36 ± 0.16	1.15	27.93 ± 0.34	2.37	0.76	0.49
5	28	20.36 ± 0.17	0.92	28.29 ± 0.41	2.16	0.63	0.43
Old	9	20.03 ± 0.24	0.71	28.59 ± 0.67	2.00	0.18	0.4

[a]The material is cumulative from old to young; the 1-year group includes the entire sample (original material).
[b] = slope of major axis.

NOTES

1. Van Valen (1962) showed partial correlations between teeth to be negative for constant tooth row length. This is of course to be expected, but in some instances the correlation was in excess of the expected value, indicating some negative interaction.

2. The original publication (Kurtén 1955b) also included a fossil *U. arctos* from Malarnaud on data from Couturier, but this specimen is now regarded as *U. spelaeus*.

REFERENCES

Bader, R. S. and J. S. Hall. 1960. Osteometric Variation and Function in Bats, *Evolution*, 14:8–17.

Degerbøl, M. 1933. Danmarks pattedyr i fortiden i sammenligning med recente former (With summary in English.) *Vidensk. Med. Densk Natur. Hist. Foren.* 96:353–641.

Flower, S. S. 1931. Contributions to our Knowledge of the Duration of Life in Vertebrate Animals. V. Mammals. *Proc. Zool. Soc.*, London.

Kurtén, B. 1953. On the Variation and Population Dynamics of Fossil and Recent Mammal Populations, *Acta Zool. Fennica* 76:1–122.

Kurtén, B. 1954. Observations on Allometry in Mammalian Dentition; Its Interpretation and Evolutionary Significance, *Acta Zool. Fennica* 85:1–13.

Kurtén, B. 1955a. Sex Dimorphism and Size Trends in the Cave bear, Ursus spelaeus Rosenmuller and Heinroth. *Acta Zool. Fennica* 90:1–48.

Kurtén, B. 1955b. Contribution to the History of a Mutation During 1,000,000 Years, *Evolution* 9:107–18.

Kurtén, B. 1957a. Life and Death of the Pleistocene Cave Bear, A Study in Paleoecology, *Acta Zool. Fennica* 95:1–59.

Kurtén, B. 1957b. A Case of Darwinian Selection in Bears, *Evolution* 11:412–16.

Kurtén, B. 1959. Rates of Evolution in Fossil Mammals, *Cold Spring Harbor Symp. Swant. Biol.* 24:204–15.

Kurtén, B. 1964. Population Structure in Paleoecology. In J. Imbrie and N. Newall, eds. *Approaches to Paleoecology.* p. 91–106. New York: Wiley.

Kurtén, B. 1965. The Carnivora of the Palestine caves, *Acta Zool. Fennica* 107:1–74.

Nilsson, E. 1953. Om sodra Sveriges senkvartara historia, (English summary) *Geol. Foren. Stockholm Forhaudl.*, 75:155–246.

Nilsson, E. 1960. Sodra Sverige i senglacial tid, *Geol. Foren. Stockholm Forhandl.* 82:134–39.

Rensch, B. 1954. *Neucre Probleme der Abstammungslchre, Die transspezifische Evolution,* Ferdinand Enke, Stuttgart.

Van Valen, L. 1962. Growth Fields in the Dentition of *Peromyscus, Evolution* 16:272–77.

FOUR

The Evolution of the Polar Bear, Ursus Maritimus Phipps

INTRODUCTION

FOSSIL REMAINS OF the polar bear, *Ursus maritimus* Phipps, are very scarce. Why this should be so is not altogether clear. In the beach deposits within its present range, for instance at Spitzbergen, polar bear bones of recent or sub-recent age are certainly quite common. There is now good evidence that the polar bear existed in Europe during the last glaciation; of course, the beach deposits from that stage are now located below sea level in most areas. Recorded finds come mostly from regions of postglacial uplift, and date from the glacial retreat stages.

The purpose of the present paper is to add some new information on the fossil occurrence of the species. One of the specimens described is a fossil of undoubted Pleistocene age, typifying a new subspecies of *Ursus maritimus*. This shows that the polar bear, like many other mammals, has decreased in size since Pleistocene times.

Apart from this, the fossil and subfossil finds give little information on the descent of the species, although there is a suggestion that some of the *maritimus* characters in the dentition may be of rather recent origin.

The dentition of the polar bear reflects the carnivorous habits of the species and diverges from the omnivorous dentition of most other ursids. If this is regarded as a primitive character, a long independent history of the species must be postulated. However, the study by Thenius (1953) suggests that the dental characters of the Polar Bear were acquired at a

Reprinted, with permission, from *Acta Zoologica Fennica* (1964) 108:1–26.

relatively late date, and that a descent from the brown bear stock is probable. Biometric comparison of growth patterns in the polar bear skull appear to corroborate this conclusion, as shown in the present paper, and may suggest that *Ursus maritimus* branched off from the evolving brown bear phylum at a fairly late date, perhaps during the Middle Pleistocene.

I wish to record my sincere gratitude to Dr. Johannes Lepiksaar, Göteborg, who kindly and generously supplied help and information, and permitted me to place on record a number of subfossil Polar Bear finds. I also gratefully acknowledge the invaluable aid and information given by Dr. A. J. Sutcliffe, London.

The most important recent comparative material available to me has been the large collection of *Ursus maritimus* from Greenland in the Zoological Museum of the University of Copenhagen, in the care of Dr. Braestrup. Additional specimens have been studied at the Vertebrate Department of Naturhistoriska Riksmuseet, Stockholm, headed by Dr. A. Johnels, and the Zoological Institute of Uppsala University, Custodian Dr. Å. Holm. The help and many courtesies extended to me in this connection is hereby gratefully acknowledged.

ABBREVIATIONS

The following abbreviations have been used to identify collections:

B.M., British Museum (Natural History), London.
C.N., Copenhagen Museum of Zoology.
H.G.M., Geological Museum, University of Helsinki/Helsingfors.
L.A.C.M., Los Angeles County Museum.
N.H.M.G., Natural History Museum, Göteborg.
N.R.M., Naturhistoriska Riksmuseet, Stockholm.
U.C., University of California.
U.Z.I., Zoological Institute, University of Uppsala.

Abbreviations used in tables of measurements:

a, approximate.
d, a difference.
e, estimated.
k, coefficient of allometry.
M, mean.
N, number of specimens.
O.R., observed range of variation.
P, probability.
r, coefficient of correlation.
S.D., standard deviation.
S_{yx}, standard deviation of regression of y on x.
t, Student's t.
V, coefficient of variation.

Subfossil finds of the Polar Bear
(Ursus maritimus maritimus *Phipps*)

At least two subfossil finds of the Polar Bear have been described previously, one from Denmark and one from southern Sweden.

1. The Danish specimen (Nordmann and Degerbøl 1930) is from Asdal, north of Hjorring, on the northern tip of Jutland. It was found in a secondary deposit, but may apparently be referred to the Late-Glacial. It is a left ramus of large dimensions, matched only by the largest recent specimens. The cheek teeth are also large, though not outside the observed range of variation of the living form.

2. The Swedish specimen (Holst 1902) comes from varved clay underlying peat at Kullaberg, Scania, and thus is of approximately the same age as the Danish one. It is a right femur. The dimensions of this specimen (measured from a cast in the Zoological Institute of Uppsala University) are given in table 4.1. Except for some wear of the greater trochanter, the specimen appears to be well preserved. It was found as early as 1852, but was not correctly identified until Holst made his study (1902) of the Swedish subfossil bears. As Holst notes, it is distinguished from *Ursus arctos* by its large size, and from both *U. arctos* and *U. spelaeus* (to which latter it had been erroneously referred) by the diagnostic character of the lesser trochanter; this is not visible in the anterior view of the *Ursus maritimus* femur. It may be added that the specimen also differs from the femur of the cave bear by its short and thick collum and the relatively slender shaft. The specimen is figured in Holst (1902, fig. 7).

Though this femur is very large, it is yet exceeded in size by a Recent specimen in the sample available to me (see table 4.1).

Two Subfossil Polar Bear Skulls from Sweden

In the Natural History Museum of Göteborg there are two skulls of the polar bear, both of them from Yoldia Clay, and thus roughly contemporary with the two specimens already mentioned. The skulls were briefly mentioned in Kurtén (1958).

3. One of the specimens, N.H.M.G. No. 1919–3356 (Coll. an. 4022), is a left upper jaw, found at Hisingen just north of Göteborg at the crossing

Table 4.1.
Measurements of bear femora.

	Length	Head Diameter	Shaft Width	Distal Width
Ursus maritimus				
Subfossil, Kullaberg	496	61	44.7	109
U.Z.I. 3 ♂	507	65	42	107
N.R.M. Novaya Zembla ♂	477	60.6	44.2	109
N.R.M. "Polar Exp." ♂	476	58	41.9	99
C.N. 1465 Greenland ♂	441	61	37.7	104
C.N. 2613 Scoresby Sound ♂	418	53.4	33.0	91
C.N. 580 Greenland ♂	405	59	34.1	103
N.R.M. ♀	403	52.1	33.2	90
C.N. 1965 Scoresby Sound ♀	385	50.6	29.6	85
N.R.M. Skansen Zoo ♀	379	50.0	31.5	84
Ursus spelaeus				
N.R.M: Jerzmanovice ♂	466	62.5	48.3	113
Coll. Koby St-Brais II ♂[1]	443	59.5	52.4	109
Ursus arctos				
U.Z.I. Lapland 1859	383	45.2	29.2	78
Subfossil, Kams Mose, Denmark[2]	450	56	38.0	93

[1]Private collection of Dr. F.-Ed. Koby, Basel.
[2]From Degerbøl, 1933.

of two streets, Bäckedalsgatan and Andersgårdsgatan. This specimen (plate 4.1, C) represents a large individual, evidently a male. The size of the canine tooth and the length of the palate (table 4.2) agree well with those found in Recent males. The left I^3, C, P^4, M^1 and M^2 are preserved; there are also alveoli for the two anterior incisors, P^1, and P^3. The second premolar was lacking in this specimen. The teeth are slightly worn. Part of the zygomatic arch is preserved.

4. The other specimen, N.H.M.G. No. 1925–4354 (Coll. an. 4884) was found at Kärraberg, Veddige, in the province of Halland; this locality is close to the Kattegat coast, approximately 50 km south of Göteborg. This specimen (plate 4.1, A and B) is a nearly complete skull. It evidently belonged to a young female individual. The width of the canine tooth, 14.2 mm, is typical of recent female polar bears, and so is the skull length, 301 mm from prosthion to basion. The teeth are unworn; both the canines,

Table 4.2.
Measurements of subfossil *Ursus maritimus* teeth and skulls, compared
with data on recent adult male and female samples from Greenland.

	Recent males			Recent females			Subfossil Hisingen	Kärraberg
	N	M	S. D.	N	M	S. D.	♂	♀
C^s, width	58	16.99	1.08	38	14.20	1.16	16.9	14.2
P^4, length	63	15.85	0.65	43	14.39	0.53	13.8	13.6
P^4, width	26	9.76	0.62	24	8.88	0.52	8.7	9.9
M^1, length	64	19.56	0.90	44	18.05	0.64	19.4	19.0
M^1, width	30	15.40	0.65	24	14.00	0.60	14.5	14.8
M^2, length	66	27.38	2.01	39	24.50	1.76	26.7	22.8
M^2, width	31	15.13	0.61	26	13.56	0.41	—	13.7
Palatal length	52	190.2	8.4	31	168.8	6.4	203	165
Basal length	52	345.9	16.4	27	306.0	12.2	—	301

the right P^4–M^2, and the left M^1–M^2 are preserved, while the incisors, P^1 and P^3 of both sides, and the left P^4 are only represented by alveoli. As in the other specimen, P^2 is absent.

On the whole the skulls and teeth do not present any significant deviation from the normal pattern in *Ursus maritimus,* to which species they unquestionably belong. However, an interesting exception is found in the unusual breadth of P^4 in the Kärraberg skull, and the large size of the protocone in this tooth. Normally the protocone in the *Ursus maritimus* P^4 is comparatively small, and frequently it is altogether absent, so that the outline of the tooth is nearly rectangular, instead of triangular as in other bears. Such is the case in the Hisingen jaw (plate 4.4, B). A condition resembling that in the Kärraberg specimen does occur in some recent *Ursus maritimus,* but it is decidedly uncommon.

If the opinion is correct that the polar bear evolved from an *arctos*-like form, a broad, triangular P^4 would be the ancestral condition. This condition might then be relatively more frequent in early populations of the evolving phylum. The fact that it is represented in a sample of only two specimens from the Yoldia Clay, when it certainly has a much lower frequency than 50 percent in the present-day form, may suggest that the frequency of this character has been changing in Postglacial time.

This may be further evaluated by means of an allometry analysis of the breadth and length of P^4 in Polar Bears and related forms (fig. 4.1). The figure shows the scatter for a sample of recent *Ursus maritimus,* mostly Greenland specimens in the Copenhagen collection. The position of the two subfossil specimens is also indicated, and in addition there are trend axes for *Ursus maritimus, U. arctos,* and the Middle Pleistocene *U. deningeri.* Mostly, in the recent Polar Bears, the left and right P^4 were rather similar, and then only one was measured; but in five specimens there are obvious differences in breadth, and in one case also in length; these were measured separately, and the corresponding symbols are joined by lines in the diagram.

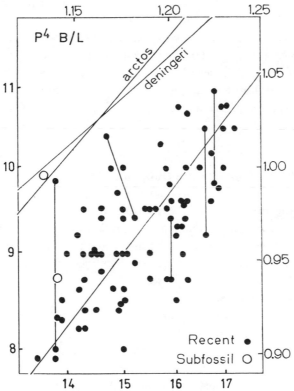

Fig. 4.1. Allometric relationships between breadth and length of upper carnassial in *Ursus maritimus,* recent and subfossil as labelled, with trend axis for the recent sample. Also trend axes for *U. arctos* and *U. deningeri,* as labelled. Measurements in millimeters along left and bottom margins; logarithms along right and upper margins.

In this diagram, the Kärraberg specimen lies entirely outside the recorded variation range in the living form, and close to the trend axes for *U. arctos* and *U. deningeri*. Only two specimens out of a total of 80 measured recent *maritimus* P^4 show conditions approaching that of the subfossil tooth, and none of these is as extreme. The second subfossil P^4, although lacking its protocone, also deviates in the same direction, but this deviation is slight and the specimen is well inside the range of the recent population.

Could the small subfossil sample of two specimens have been drawn from a population similar to the recent? Analysis of variance gives P = 0.002, so that the odds are 500 to 1 against drawing a sample of this kind from the living population.

It may be concluded that the Late-Glacial polar bear had diverged less from the *arctos* type, as far as the relative width of P^4 is concerned, than the living form. In other words, the present-day type of P^4 in *Ursus maritimus* is a geologically late acquisition, and its evolution was still under way about ten thousand years ago, and is probably still going on.

Subfossil Polar Bear remains from Uddevalla

From Dr. Johannes Lepiksaar of the Natural History Museum of Göteborg I have received the following information on some additional subfossil specimens in the collections of the Museum, all of which are from the shellbanks at Uddevalla. They have been identified by Dr. Lepiksaar.

5. Bräcke, April 1932 (No. 1932–5684, Coll. an. 6003), distal half of the diaphyse of a left humerus.

6. Kuröd (No. 1931–5494, Coll. an. 6020), fragment of right scapula.

7. Bräcke (No. 1936–6990, Coll. an. 6565), left upper canine. The specimen in the Museum is a cast, and the original is the property of Captain T. Hyltén-Cavallius, Göteborg. The specimen was published by L. A. Jägerskiöld in the 1936 Annual Report of the Museum, pp. 9–10.

8. Nedre Kuröd (No. 1952–9085, Coll. an. 7526–7528) three proximal fragments of the second, fifth, and sixth ribs of the right side.

PLEISTOCENE FINDS OF THE POLAR BEAR
(*Ursus maritimus tyrannus,* new subspecies)

Description

Type: A right ulna, B.M. No. 24361. See Plate 4.2, A.

Type locality and horizon: New Kew Bridge, London; Late Pleistocene, probably Early Würm.

Diagnosis: A polar bear greatly exceeding the living form in size; length of type ulna about 485 mm.

1. There seems to be only one reliable previous record of the polar bear in Pleistocene deposits. This is a skull which was mentioned by Zimmermann (1845) as found in Hamburg during sewage constructions (see Erdbrink, 1953). The skull was not described and its present whereabouts are unknown. However, the fact that it was found together with bones of the Walrus seems, as Erdbrink remarks, to support the determination as *Ursus maritimus.* The specimen apparently dates from the Würm. There is no information on the size of this specimen.

2. An alleged find (Tscherski, 1892) from the Bolshoi Island is rejected by Erdbrink (1953). The specimen in question was an isolated canine tooth, occurring with a Late Pleistocene fauna also including brown bear. Under such circumstances the determination would seem to be hazardous indeed, though, of course, it may be correct.

3. There is an alleged find from Alaska (see Hay 1937), but its Pleistocene age seems to be in doubt.

4. The ulna from Kew Bridge, the type of *Ursus maritimus tyrannus,* is the only specimen that at present will throw any light on the Pleistocene history of this species. Although this is only a single specimen, study and comparison shows (a) that it does not belong to any other known species of bear than *Ursus maritimus,* and (b) that it must have belonged to a form greatly exceeding the living polar bear in size.

Bear taxonomy, especially as regards the brown and cave bears, is in a state of great confusion, because of the creation of an immense number of superfluous taxa. It follows that the greatest caution should be exercised in naming new forms. For this reason I have thought it useful to give a fairly detailed account of the morphological and biometric study of the London specimen. As all the Middle and Late Pleistocene large bears of Europe are considered in this comparison, the results may be of some use to other students faced with determination problems.

There is no sharp boundary between morphological and numerical analysis. Some characters are easy to express numerically, others are not. The latter are treated in the section headed "morphological comparison", the others form the topic of the "numerical comparison."

B.M. No. 24361 is a bear ulna of gigantic proportions. The specimen is well preserved, but the distal epiphyse is missing, so that the bone evidently belonged to a subadult animal. There is also slight damage to the distal end of the diaphyse, the olecranon, and the coronoid process, but the essential characters are well preserved. The fragment has a total length of about 440 mm as preserved. In a very large subadult recent polar bear (U.Z.I. No. 3), the length without distal epiphyse is 387 mm, and the total length 428 mm. Thus the full length of the fossil bone may be estimated at a minimum of 485 mm. As far as I know this is the longest ursine ulna on record. The large Pleistocene tremarctines *(Arctodus)* of the Americas were unusually long-legged bears, and in some cases their limb bones attain similar or greater dimensions. The maximum record from Potter Creek Cave in northern California is 446 mm (U.C. No. 3426), and from Rancho La Brea, Los Angeles, 475 mm (L.A.C.M. No. Z32); and a tremendous specimen from the Irvingtonian, or Californian Middle Pleistocene, attains the almost incredible length of 591 mm.

It is, of course, *a priori* unlikely that an arctodont bear would have existed in Europe during the Pleistocene, and the morphological characters of the Kew Bridge bone in fact exclude the possibility of such affinities, but for the sake of completeness data on an arctodont ulna (from Potter Creek Cave) have been included in the comparisons.

Morphological Comparison

Plates 4.2 and 4.3 show a series of bear ulnae, all to the same scale. In spite of some damage in this part of the Kew Bridge specimen, it can still be seen that the olecranon process rose above the semilunar notch in a way that serves to distinguish the bone from its homologue in *Ursus deningeri* and *Ursus spelaeus*. Furthermore, the dorsal margin is nearly straight, instead of concave as in these two forms. In *U. arctos* it is variable, but modally somewhat concave. In both characters the Kew bone is indistinguishable from *U. maritimus*. The shape of the olecranon accords more closely with the polar bear than with any other form. The articulation facet of the semilunar notch, in its middle part, merges with the medial side of the bone, as in *U. maritimus;* in *U. arctos* the boundary is more conspicuous.

The internal profile of the semilunar notch forms an almost perfect semicircle in the polar bear; at the lower end it continues into a short, straight, tangential edge at right angles to the long axis of the bone. The characters in *U. arctos* are nearly similar. In *U. deningeri* the Cavum sigmoideum faces more dorsad and the coronoid process is thus more prominent. The notch of *U. spelaeus* is not semicircular, as the upper side of the coronoid process curves downward. The apex of the coronoid process is blunt or rounded in *U. arctos* and the spelaeid bears *(U. spelaeus, U. deningeri)* but tends in the polar bear to be produced downward into a point. The base of this structure is seen in the Kew specimen, but the point itself is broken.

The shaft tapers very gradually towards the distal end, as in the polar bear, and the course and development of the attachment area for the interosseous ligament are similar. In *U. arctos* the taper is more pronounced. The shaft is straight as in polar bear and brown bear; in the spelaeids it has a sigmoid curvature, the dorsal margin being concave proximally and convex distally, where there is a sudden bend just above the capitulum.

Characters of size and relative proportions may also be evaluated by direct morphological comparison, but a numerical study is more useful.

Numerical Comparison

The following measurements, singly or in combinations, were found to give good characters for the separation of the taxa which have been compared here (see table 4.3).

L, total length of the ulna.

PD, greatest proximal diameter, measured anteroposteriorly from the tip of the coronoid process to Margo dorsalis.

SD, inner diameter of the semilunar notch, measured vertically.

BS, minimum transverse breadth of shaft. This is in the distal portion of the shaft, at some distance above the capitulum. The extreme constriction lies close to the capitulum in the brown bear and the spelaeids, but higher on the shaft in the polar bears, including the Kew specimen.

A comparison in the form of a ratio diagram (fig.4.2), in which the Kew Bridge ulna is taken as the standard, gives an idea of relative proportions. Two specimens of *Ursus maritimus,* the largest male (plate 4.2, B) and the smallest female, give patterns tending to the straight vertical, showing that their relative proportions are nearly the same as those of the fossil bone. In both of them the shaft is relatively plumper than in the ulna from Kew, but this difference is slight, and two other recent specimens (C.N.1965 and C.N.580) are almost exactly like the fossil bone regarding the relationship between BS and L. On the other hand, the shaft is relatively plumper than in *U. arctos*.

All the other bones give more or less aberrant patterns, and the results may be summarized as follows.

Ursus arctos, Late Pleistocene and recent. A fossil specimen from Crayford, a recent from Alaska, and a recent from Sweden; the last-mentioned is too small to be depicted in the ratio diagram. The Crayford ulna differs from the Kew specimen in the greater constriction of the shaft; this also holds for the Alaska bear, which in addition has a larger semilunar notch.

Ursus deningeri and *U. spelaeus.* The former is represented by an ulna from the Bacton Forest Bed, the latter by a specimen from Odessa. As regards L, PD, and SD the two form a consistent pattern which deviates strongly from the brown and polar bears. The proximal diameter is rela-

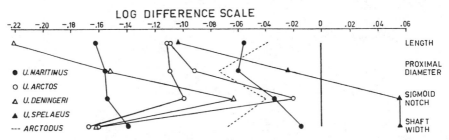

Fig. 4.2. Ratio diagram, comparing the type ulna of *Ursus maritimus tyrannus* (standard, straight vertical at o) with various other bear ulnae as labelled.

tively long and the semilunar notch very large. Finally, in both forms the shaft breadth is much greater, relative to ulnar length, than in the other bears. This condition is more extreme in the cave bear than in *U. deningeri*. It may be noted that these characters will serve very well for distinguishing between the ulnae of *U. spelaeus, U. deningeri,* and *U. arctos priscus,* the most common Pleistocene bears in Europe.

In *Arctodus* the proportions as regards these measurements do not deviate much from *U. arctos* or *U. maritimus*. The ulna of the American short-faced bear, however, is readily distinguished from *Ursus* on morphological characters (Merriam and Stock 1925:p.24).

The absolute length of the Kew Bridge ulna also is an important character, when compared with the lengths reached by other bear ulnae. Ulnae of *Ursus deningeri* reach lengths of 320 mm (Hundsheim) and 325 mm (Mosbach; both records from Zapfe 1946). None of the British Museum specimens from the Forest Bed attain quite as large dimensions. In *Ursus spelaeus* much longer ulnae may be found. Out of the very large material known, maximum dimensions of 420 and 441 mm are cited by Koby (1950). The latter record is probably rather close to the real population limit, as these extremes are based on very large samples. Of course, this is still far from the 485 mm of the Kew Bridge bone.

The ulna of the brown bear may attain very large dimensions. Among these may be cited 377 mm (Alaska), 376 (Pleistocene, Crayford); 367 (recent Russian in Borissiak 1932); and 357 mm (recent European in Ehrenberg 1942). Some Pleistocene brown bears probably attained much greater size. A proximal fragment of an ulna from Brixham (B.M. No. 48832) probably represents a specimen of more than 400 mm in length.

It would thus appear that lengths of more than 400 mm are fairly unusual in *U. spelaeus* (attained only by the largest males); very rare in *Ursus arctos priscus* and other forms of brown bear; and unknown, as well as improbable, in *U. deningeri*. In *U. maritimus,* on the other hand, such dimensions are a commonplace. Of six male individuals measured, two have ulnae longer than 400 mm.

It may thus be concluded that the ulna from Kew Bridge must be referred to the species *Ursus maritimus,* of which it represents a very large form. The distinction in size between this form and the living population will be considered in the following section.

Differentiation in Size

The Kew Bridge ulna is without doubt much too large to fit into the distribution of recent *Ursus maritimus*. The results of a statistical comparison between the measurements of the fossil bone and those of the recent sample of table 4.3 are set down in table 4.4. The differentiation in total length is highly significant, that in the proximal diameter is also probably significant, whereas the diameter of the sigmoid notch and the shaft width are not significantly different from the means for the recent sample.

It is often said that much of the differentiation between fossil and

Table 4.3.
Measurements of bear ulnae.

	L	PD	SD	BS[a]
Ursus maritimus				
B.M. 24361 Kew Bridge	e485	a95	44	27.6
U.Z.I. 3 ♂	428	83	40.7	26.7
N.R.M. Novaya Zembla ♂	410	84	38.6	28.0
N.R.M. "Polar Exp." ♂	395	80.3	37.6	25.4
C.N. 1465 Greenland ♂	386	74	41.0	23.9
C.N. 2613 Scoresby Sound ♂	364	67	35.0	21.4
C.N. 580 Greenland ♂	362	71	35.7	20.8
N.R.M. ♀	350	69	32.7	22.6
C.N. 1965 Scoresby Sound ♀	339	62	30.4	19.5
N.R.M. Skansen Zoo ♀	334	66.5	30.9	20.3
Ursus arctos				
B.M.M. 9603 Crayford	376	90	50	21.4
N.R.M. 233 Sweden	281	47.0	30	13.7
H.Z.M. Alaska	377	77	42	19.1
Ursus deningeri				
B.M. 10500 Bacton	292	67	38	19.0
Ursus spelaeus				
H.G.M. Odessa ♂	383	90	50	31.4
Arctodus sp.				
U.C. 3426 Potter Creek Cave	445	78	40.1	23.6

[a]L, length; PD, maximum proximal diameter; SD, inner diameter of semilunar notch; BS, minimum shaft diameter.

Table 4.4.
Comparison between dimensions in fossil ulna of *Ursus maritimus* and means of recent sample.

	N	M	S.D.	V	d	t	P[1]
L	9	374.2 ± 10.8	32	8.7	110.8	3.42	<0.01
PD	9	73.0 ± 2.6	7.9	10.8	22.0	2.80	~0.02
SD	9	35.8 ± 1.3	4.0	11.1	8.2	2.1	>0.05
BS	9	23.2 ± 1.0	3.0	12.9	4.4	1.5	>0.10

[1]d, difference between measurement of fossil bone and mean for recent sample; t, d/S.D.; P, probability.

subfossil mammals on one hand, and their recent descendants on the other, is simply because most individuals are now killed before they attain fully adult size. However, this explanation does not fit the present case, as the fossil bone represents a subadult individual, in spite of its immense size. Incidentally, the "explanation" does not explain away the instances in which the differentiation is based on dental measurements that do not change with age. I have elsewhere (Kurtén 1964) suggested another, seemingly more important factor in the Postglacial dwarfing of so many mammalian species.

The possibility of spatial differentiation in size in the present-day population of the polar bear should not be overlooked. The Copenhagen sample is from Greenland. Ognev (1935) has published the measurements of a number of male skulls of the alleged subspecies *U. maritimus marinus* from the Eurasian north coast, that is from the opposite part of the range of the species. Table 4.5 shows a comparison between this sample and the adult males from Greenland. Evidently there is no real differentiation between the western and eastern populations of polar bear at the present time.

Geological Age of the Polar Bear from Kew

The fossil ulna comes from a railway cutting near Kew Bridge and thus is likely to be close in age to the so-called Brentford fauna, which dates from the end of the last interglacial (Zeuner 1958). The fauna, which contains the species *Crocuta crocuta* (Erxleben), *Hippopotamus* sp., *Cervus*

Table 4.5.
**Comparison between skull measurements of adult male *Ursus maritimus*
from Greenland and Eurasia.**

	N	O.R.	M	S.D.	V	P
Basal Length						
Greenland	52	311–389	345.9 ± 2.3	16.4	4.7	} >0.10
Eurasia	8	328–360	341.9 ± 3.7	10.3	3.0	
Zygomatic Width						
Greenland	10	190–255	221.6 ± 7.2	22.6	10.2	} >0.10
Eurasia	11	189–274	229.3 ± 7.4	24.7	10.7	
Rostral Width						
Greenland	53	79–101	92.2 ± 0.7	5.4	5.9	} >0.10
Eurasia	11	76.3–103.0	92.8 ± 2.2	7.2	7.7	
Interorbital Width						
Greenland	17	79–105	92.4 ± 1.7	7.1	7.7	} >0.10
Eurasia	11	75.7–104.3	94.8 ± 2.4	8.1	8.5	

elaphus L., *Megaloceros* sp., *Bos primigenius* Bojanus, *Bison uniformis* Hilzheimer, and *Elephas antiquus* Falconer, is an interglacial forest fauna. The Polar Bear would be an odd intruder in this assemblage. However, Dr. A. J. Sutcliffe has informed me that the collections of the British Museum (Natural History) also contain a lot of hitherto undescribed reindeer fossils from Kew, supporting the inference that there is also a cold fauna here. Perhaps the most probable interpretation at present would be that the sequence at Kew covers both the end of the Eemian Interglacial and the beginning of the Early Würm. Thus the date of *Ursus maritimus tyrannus* may be tentatively set at Early Würm.

THE ANCESTRY OF THE POLAR BEAR

The available fossil material does not give any direct evidence of the descent of the polar bear except for the suggestion that the *maritimus* type of P^4 may have evolved at a comparatively late date from a more *arctos*-like form. Indirectly, a relatively late origin may also be suggested by the absence of polar bear fossils in deposits antedating the Würm. On the other hand, the fact that *Ursus arctos* and *Ursus maritimus* may produce fertile hybrids also suggests a fairly close relationship between the two species (see discussion and references in Thenius 1953).

Additional evidence may be found by comparison between the anatomy of the polar bear and that of other fossil and recent bears. One of the important points is the evidence suggesting a reduction in size of the teeth, especially M^2. A study of the allometric growth patterns of the polar bear skull may also throw some light on this question.

Reduction of the Second Upper Molar

The postcarnassial teeth of the polar bear are relatively small, whereas the carnassials have high, sharply pointed cusps. This is related to the carnivorous adaptation of the species, and deviates sharply from the normal trend of evolution in *Ursus*. That the second upper molar has in fact been reduced is clearly shown by the root system of this tooth, as was shown by Thenius (1953). Though the polar bear M^2 has a relatively short talon, usually much narrower than the trigon, the root system nonetheless is of the same type as that of *Ursus arctos*. The well-developed inner root, which resembles that of the brown bears, is a striking character. In the black bears, which never achieved the same elongation of M^2 as the brown bear and cave bear group, this root is relatively feeble. The condition in *Ursus maritimus*, as Thenius concludes, points back to an ancestral stage possessing a large talon in M^2.

That the talon of M^2 is actually in the process of being reduced is also, if not proved, at least suggested by the variation of this tooth as exemplified in the Greenland collection in Copenhagen. The length of the molar is exceedingly variable (table 4.6), which may suggest loss of adaptive importance. For instance, 66 male individuals give a coefficient of variation $V = 7.35$; 39 females, $V = 7.18$. The combined sample (which also includes 8 unsexed juvenile specimens) has $V = 9.62$.

A sample of recent *Ursus arctos,* both sexes combined, from Fennoscandia, (47 Swedish and 46 Finnish specimens on original data, and 25 Norwegian on data from Degerbøl 1933) has a V of 7.05. This value should be compared with that for the combined polar bear material. The difference

Table 4.6.
Variation in the length of M^2 in ursids.

	N	O.R.	M	S.D.	V
U. maritimus ♂ ♂	66	21.6–30.8	27.38 ± 0.25	2.01	7.35
U. maritimus ♀ ♀	39	20.1–27.4	24.50 ± 0.28	1.76	7.18
U. maritimus ♂ ♀	113	18.7–30.8	26.18 ± 0.24	2.52	9.62
U. arctos ♂ ♀ Fennoscandia	118	26.7–39.1	31.62 ± 0.21	2.23	7.05
U. spelaeus ♂ ♀ Odessa	93	39.2–52.0	46.28 ± 0.28	2.66	5.74
U. spelaeus ♂ ♂ Mixnitz	44	43.4–52.5	48.44 ± 0.33	2.21	4.57

between the two is highly significant (P<0.01). A sample of 93 unsexed M^2 of *Ursus spelaeus* from Odessa has a still lower V, 5.74; comparison with the polar bear value gives t = 5.07 and P<0.001.

Finally, a sexed sample of *Ursus spelaeus* (Mixnitz males, V = 4.57) may be compared with the corresponding polar bear sample (*U. maritimus* males, V = 7.35). This difference is also significant (P<0.01). There is no doubt that the polar bear M^2 is more variable in length than its homologues in the brown bear and cave bear.

A study of the frequency distributions (table 4.7) is also illuminating. They are markedly skewed, with modes in the upper part of the range, and a long "tail" of low values. Some of these length measurements appear almost incredibly small. They represent teeth in which the talon has vanished altogether, and the crown, consisting virtually of the trigon only, is a nearly circular structure (Plate 4.4, A). The shortest M^2 (in C.N. No. 1859, an unsexed juvenile specimen) has a length of 18.7 and a breadth of 13.3 mm.

The evidence thus appears to point back to an ancestral form of orthodox bear type, probably of the general brown bear stock in the Lower or Middle Pleistocene. A black bear ancestry seems to be negated by the characters in the M^2 roots, while on the other hands the cave bear line probably was too specialized to figure in this connection.

Allometric Growth Patterns in the Polar Bear Skull

Some circumstantial evidence may be obtained from a study of the allometry characters in polar bears and various living and extinct ursids.

Table 4.7.
Distribution of lengths of M^2 in *Ursus maritimus*.

Length Class	Frequencies ♂♂	♀♀	♂♀
18.3–19.2	—	—	1
19.3–20.2	—	1	1
20.3–21.2	—	2	3
21.3–22.2	1	1	2
22.3–23.2	1	3	5
23.3–24.2	1	9	11
24.3–25.2	9	12	22
25.3–26.2	7	4	11
26.3–27.2	11	5	16
27.3–28.2	9	2	13
28.3–29.2	16	—	16
29.3–30.2	8	—	9
30.3–31.2	3	—	3
Totals	66	39	113

Some data of this type will be presented and discussed here. They are summarized in table 4.8.

ALLOMETRY OF MUZZLE BREADTH
AND BASAL LENGTH OF SKULL

The muzzle breadth has been measured between the external faces of M^2, the skull length from prosthion to basion. The results are set forth as an allometry scattergram in fig. 4.3.

In the living and Late Pleistocene brown bears the growth axis shows a distinct and sharply localized change in allometry. In youth and up to a certain size, the breadth grows slowly in relation to the length, so that the allometry is negative; k, the coefficient of allometry, is 0.41. At a length of about 260–300 mm, however, it changes and becomes nearly isometric ($k = 0.97$). Incidentally, this will lead to very complex and *a priori* unpredictable trends in breadth/length indices, so that such figures (as always in the study of allometric characters) will be useless and misleading.

A similar change (from $k = 0.52$ to $k = 1.06$) occurs in the Polar Bear, but at a somewhat later stage in the growth of the skull length (about 330–360 mm.). Moreover, the *U. maritimus* trend lines, in both phases, run below those for *U. arctos*, so that at any stage in growth a polar bear

Table 4.8.
Allometric growth patterns in bear skulls and jaws.

	N	M_{logy}	M_{logx}	r	k	S_{yx}
y = muzzle breadth, x = basal length						
U. maritimus, phase 1	47	1.8741	2.4534	0.73	0.52	0.0178
U. maritimus, phase 2	47	1.9200	2.5394	0.71	1.06	0.0151
U. arctos, phase 1	57	1.8957	2.4257	0.78	0.41	0.0173
U. arctos, phase 2	54	1.9341	2.4812	0.893	0.97	0.0235
arctos-like spelaeids	32	2.1479	2.5983	0.84	0.96	0.0226
maritimus-like spelaeids	7	2.1202	2.6345	0.85	0.90	0.0159
y = rostral breadth, x = basal length						
U. maritimus	101	1.9314	2.5107	0.913	1.18	0.0187
U. arctos (Kamtch., Kolym)	20	1.9420	2.5145	0.919	1.36	0.0173
U. arctos (various Asiatic)	41	1.9101	2.5066	0.845	0.91	0.0190
U. deningeri	8	1.9845	2.6008	0.81	0.86	0.0227
U. spelaeus female	18	1.9645	2.5717	0.872	1.15	0.0111
U. spelaeus male	41	2.0548	2.6360	0.702	1.72	0.0214
y = total length, x = basal length						
U. maritimus	11	2.5645	2.5190	0.970	1.021	0.0072
U. deningeri	6	2.688	2.628	0.985	1.01	0.010
U. arctos	93	2.4996	2.4476	0.975	1.083	0.0107
U. spelaeus	67	2.6681	2.6155	0.984	1.068	0.0092
y = palatal length, x = basal length						
U. maritimus	99	2.2403	2.4961	0.987	0.90	0.0070
U. etruscus & U. deningeri	7	2.3245	2.5530	0.989	0.99	0.0098
U. arctos	63	2.1890	2.4339	0.957	0.95	0.0111
U. spelaeus	35	2.3914	2.6076	0.957	1.11	0.0123
y = coronoid height, x = mandible length						
U. maritimus	26	1.9653	2.3499	0.949	1.27	0.0200
U. deningeri, Mosbach	8	2.0833	2.4345	0.79	1.31	0.0195
U. "deningeri," Bacton	6	2.1028	2.3928	0.90	1.17	0.0211
U. arctos	60	1.9560	2.3477	0.919	1.24	0.0211
U. spelaeus	44	2.1161	2.4620	0.926	1.20	0.025

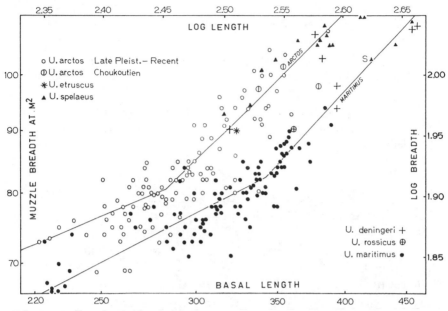

Fig. 4.3. Allometric relationships between muzzle breadth and basal length in the skull of bears, as labelled. Growth axes represent *U. arctos* and *U. maritimus*, S, *Ursus spelaeus* from Swanscombe.

skull will, on an average, be narrower across M^2–M^2 than a brown bear skull of the same length. (There is a slight overlap of variation in this respect, but the trend lines are markedly and significantly different in both phases of growth.)

Of the fossil bears, *Ursus etruscus* is comparable to the *arctos* pattern. On the other hand, data on the Cromer-Mindel *Ursus deningeri* fall into two distinct groups: one small series of specimens clusters about the *arctos* axis, the other follows the *maritimus* axis. Here is also found the eastern type of *U. deningeri*, *U. "rossicus"* from Krasnodar. The same dichotomy is seen in the Late Pleistocene descendant of *U. deningeri,* viz., the true cave bear, *U. spelaeus.* Although most specimens are of *arctos* type in this respect, a small group is clearly separated, and lies along the *maritimus* axis; this latter group includes among others the earliest known skull of *U. spelaeus,* that from Swanscombe (Kurtén 1959; S in the figure).

Table 4.9 shows the distribution into "arctoid" and "maritimoid" individuals in local populations of *U. deningeri* and *U. spelaeus*. Both types evidently may occur in the same population and in either sex, but the incidence of the "m-type" is much higher in *U. deningeri* (57%) than in *U. spelaeus* (9%) although it probably was high in the earliest true Cave Bear, judging from the Swanscombe specimen.

It would appear that the "m-type" was fairly common in the Brown and Cave Bear group in the earlier Middle Pleistocene (the Cromer and Mindel).

ALLOMETRY OF ROSTRAL BREADTH
AND BASAL LENGTH OF SKULL.

The rostral breadth is measured at the canines. The results are summarized in fig. 4.4.

Local recent populations of *Ursus arctos* show some differentiation in this character; the various trend axes that I have been able to establish differ significantly from each other as regards slope or position. There is no suggestion of a break or change in relative growth rate as in the previous case. A group from the eastern part of Siberia, probably equivalent to the subspecies *U. arctos beringianus* (from Kamtchatka and the Kolym Peninsula) is characterized by relatively strong positive allometry (k = 1.36) and an isolated position (relatively broad rostrum). The trend line of *U. maritimus* seems to be quite similar to that in *beringianus*,

Table 4.9.
Distribution of allometry patterns (y = muzzle breadth, x = basal length) in populations of *Ursus deningeri* and *U. spelaeus*.

	Locality	Sex	"a-type"	"m-type"
U. deningeri:	Mosbach	♂	—	2
	Mosbach	♀	2	1
	Bacton	♂	—	1
	Hundsheim	♀	1	—
U. spelaeus:	Swanscombe	♂	—	1
	Merkenstein	♂	2	1
	Sundwig	♂	1	1
	Various	♂♀	26	—

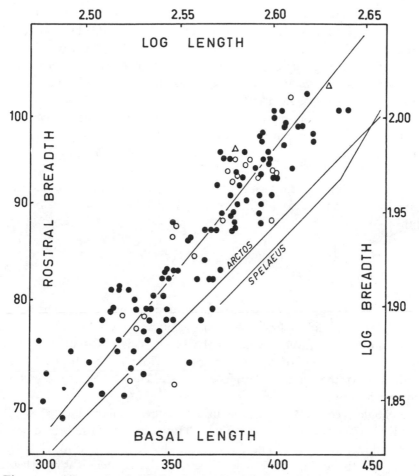

Fig. 4.4. Allometric relationships between rostral breadth and basal length in the skull of bears. Filled circles, *U. maritimus;* open circles, *U. arctos beringianus;* triangles, *U. arctos,* Choukoutien. Trend line for *U. maritimus;* also trend lines for another recent population of *U. arctos,* and for *U. spelaeus.*

and certainly differs from all other *arctos* axes found in the recent population.

In this connection it is of interest to trace the history of these growth patterns. Nearly all the subfossil and fossil samples of *U. arctos* known to me, as well as *U. deningeri,* fit in with some of the other recent *arctos* patterns, and differ significantly from the *beringianus-maritimus* group.

Even *U. spelaeus* appears to "start out" on one of these *arctos* trend lines, perhaps the same as *deningeri*, though in larger individuals a different growth pattern with very strong positive allometry is formed.

Only one small group of fossil bears appears to conform to the aberrant *beringianus-maritimus* type: two specimens from Choukoutien, dating from the earlier Middle Pleistocene (Mindel). Since the Choukoutien Brown Bear probably represents the direct ancestry of the present-day eastern Asiatic local population, this is hardly surprising, but it is of some interest to see that a differentiation in this respect had appeared at such an early date.

Again, this character may suggest some kind of connexion between the polar bear on one hand, and the Cromer-Mindel brown bear stock on the other.

ALLOMETRY OF TOTAL SKULL LENGTH AND BASAL LENGTH.

The total length is measured from prosthion to inion. In this character the Polar Bear differs from late Pleistocene and recent *U. arctos;* the total length is relatively greater in the latter. This is mainly due to a greater development of the overhanging occiput in the brown bears, and also to their greater height of skull. The difference is highly significant.

A single record for *U. etruscus* may fit either pattern, and the same holds for the Choukoutien brown bear. However, one fossil brown bear population gives a trend line not distinguishable from that in *U. maritimus;* this is, again, the Middle Pleistocene *U. deningeri*. While these differ significantly from *U. arctos,* no difference at all could be found in the comparison *deningeri-maritimus* (t less than 1).

Ursus spelaeus is somewhat closer to *U. arctos;* it differs from *deningeri* (t = 3.22), but also from *arctos,* and the axis has an intermediate position.

ALLOMETRY OF PALATAL LENGTH AND BASAL LENGTH OF SKULL.

The palatal length was measured from prosthion to staphylion. The usual pattern found was negative allometry, but in the cave bear this relationship is positively allometric. The cave bear also has the relatively longest palate. *Ursus deningeri* is intermediate between the cave bear and the later brown bears. In the Polar Bear the palate is slightly shorter than in *U. arctos,* and I have not been able to match this pattern in any other bear population.

ALLOMETRY OF THE HEIGHT OF THE CORONOID PROCESS AND THE LENGTH OF THE LOWER JAW.

The length of the jaw was measured from the anterior face of the canine to the middle of the condyle, this being the most useful dimension as regards fossil material, in which the symphysis frequently is damaged. The height of the ascending ramus is measured from the top of the process to the lower border just in front of the angle.

All the samples exhibit a weak positive allometry. The coronoid process tends to be very slightly higher in *U. arctos* than in *U. maritimus* jaws of the same length; the overlap between individual variants is great, but the two axes differ significantly in position. In *U. spelaeus* the process tends to be still somewhat higher, and the positive allometry is weaker.

A sample of the Middle Pleistocene *U. deningeri* from Mosbach differs significantly from the Late Pleistocene and recent *U. arctos,* but not from *U. maritimus*. On the other hand, the *U. deningeri* from the Bacton Forest Bed are quite aberrant, and have a very high ascending ramus; the depth of the horizontal ramus is also unusually great. This form is probably of somewhat older date than the Mosbach bear. In the history of the brown and cave bears, thus, a *maritimus*-like pattern seems to date back again to approximately the Cromer-Mindel stage.

Allometry Comparisons and the Age of the Polar Bear

The allometric growth patterns in the Polar Bear skull and jaw have been found not to be distinguishable from patterns observed in the following recent or fossil bear samples:

Muzzle breadth/Basal length: *Ursus deningeri,* early Middle Pleistocene; also a few Late Pleistocene *U. spelaeus*.

Rostral breadth/Basal length: *Ursus arctos,* early Middle Pleistocene, China; *U. a. beringianus,* recent.

Total length/Basal length: *Ursus deningeri,* early Middle Pleistocene.

Palatal length/Basal length: Found nowhere.

Coronoid height/Jaw length: *Ursus deningeri,* Cromer-Mindel population.

Out of five patterns studied, one is unique; three may be matched with patterns found in the brown and cave bear stock in the early Middle Pleistocene only (not counting persistence in *U. spelaeus*); and one with a pattern appearing in the early Middle Pleistocene and persisting to the present in the brown bear.

Of course, such matching does not constitute positive identification. However, when we consider the bewildering variety of allometric axes to be found for any given pair of variates in different ursid populations, the similarities encountered in the present case appear to be too many to be coincidental. Everything would seem to point to some special relationship with the brown bear–incipient cave bear group that lived during the early Middle Pleistocene, and especially during the Mindel Glaciation.

At that time the situation was approximately as follows. The ancestral *U. etruscus* of the Villafranchian had given rise to two divergent evolving phyla. The cave bear phylum was localized in Europe, where its progress in a more spelaeid direction may be followed in the temporal population sequence Bacton–Hundsheim–Mauer–Mosbach "Main fauna." This branch is usually called *Ursus deningeri;* I have earlier (Kurtén 1959) suggested its inclusion as a subspecies under *U. arctos,* but the new extensive material from the type locality (Mosbach) collected in recent years indicates that there was full specific separation between the two branches by Mindel times. Probably this form could just as well be regarded as an early subspecies of *U. spelaeus,* as Ehrenberg (1928) concluded.

The Brown Bear phylum, on the other hand, was localized in Asia; unfortunately, our knowledge of this form is at present limited to a relatively meagre sample, mostly from Choukoutien.

The ancestry of the Polar Bear obviously does not lie in the Mindel-age *U. deningeri,* which had already evolved various cave bear specializations such as the loss of the anterior premolars, the more or less developed doming of the skull, and a shortening of the distal limb segments. The actual ancestral population probably lay within the Asiatic Brown Bear population of this time, perhaps in the vast Siberian area,[1] but of this hypothetical ancestor we do not know anything at present.

Population fractionation during a major glaciation, for instance the Mindel or the Riss, may have produced the isolation necessary for the Polar Bear to break the ties with the ancestral species. Once it had invaded its new habitat, the ancestral Polar Bear would be subjected to strong selection pressure. Its new mode of life is a radical departure from the general run of ursine habits and hence would probably lead to rapid evolutionary change (Hecht 1963). The evidence indicates that this change has been going on in Postglacial times and probably still is continuing, for instance the loss of the protocone in P^4 and of the talon in M^2. Apparently the trend is towards developing a series of homodont cheek teeth as in many other aquatic carnivores; the cusps of the cheek teeth are already higher and more pointed than in the brown bear. A specimen in the Copenhagen collection has P^4 duplicated into two teeth one behind the other, which suggests one possibility of further change. In the lower jaw, the loss of the third molar, which is already very small, would be a logical consequence of the reduction of the talon in M^2, and it has in fact been realized in some individuals.

Although the polar bear uses its swimming powers to move about in quest of its prey, the actual hunting always takes place on the ice or the ground, so that no radical change in the limbs would be expected; and they are in fact very little modified from the brown bear type.

It is to be hoped that future research will lead to the discovery of better fossil evidence on the evolution of the polar bear. As an example of an animal that has invaded a new adaptive zone, and probably has evolved at a fairly rapid rate, it may be expected to throw important light on the processes leading to the evolution of new grades of organization.

SUMMARY

This paper lists known fossil and subfossil material of the polar bear, *Ursus maritimus,* and describes new finds. The Late Pleistocene polar bear, *U. maritimus tyrannus,* new subspecies, was markedly larger than the living. As late as in Yoldia times, the polar bear P^4 was on an average more *arctos*-like than at present, indicating that evolution within the species has continued in the Postglacial. M^2 shows evidence of secondary reduction from an original condition resembling that in the Brown Bears.

Allometric growth patterns in the Polar Bear skull may in several instances be matched by allometries found in early Middle Pleistocene (especially Mindel) Brown and Cave Bears, suggesting that the ancestry of the Polar Bear was close to this stock.

NOTES

1. A North American origin is not probable, for as far as we know *V. arctos* did not enter North American until the last glaciation (Kurtén 1960.)

REFERENCES

Borissiak, A. 1932. Eine neue Rasse des Höhlenbären aus den quartären Ablagerungen des Nordkaukasus. Trav. Inst. *Paleozool. Acad. Sci. URSS* (for 1931)1:137–201.

Degerbøl, M. 1933. Danmarks Pattedyr i Fortiden i Sammenligning med recente Former. I. Vidensk. Med. *Dansk Maturhist. Foren.* 96:355–641. (With English summary.)

Ehrenberg, K. 1928. *Ursus deningeri* v. Reich. und *Ursus spelaeus* Rosenm. *Akad. Anz.,* vol. 10, *Akad. Wiss. Wien, math.-nat. Kl., Sitz.* 26 April 1928, pp. 1–4.

Ehrenberg, K. 1942. Berichte über Ausgrabungen in der Salzofenhöhle im Toten Gebirge. II. Untersuchungen über umfassendere Skelettfunde als Beitrag zur Frage der Form- und Grössenverschiedenheiten zwischen Braunbär und Höhlenbär. *Palaeobiol.* 7:531–666.

Hay, O. P. 1929. *Second Bibliography and Catalogue of the Fossil Vertebrata of North America.* Carnegie Inst. Washington Publ., vol. 390.

Holst, N. O. 1902. Några subfossila björnfynd. *Afhandl. Svergies Geol.* Unders., ser. C., 189:1–38.

Koby, F.-Ed. 1950. Les dimensions minima et maxima des os longs d'*Ursus spelaeus. Ecl. Geol. Helvetiae* 43:287.

Kurtén, B. 1958. The bears and hyenas of the interglacials. *Quaternaria* 4:69–81.

Kurtén, B. 1959. On the bears of the Holsteinian interglacial. *Stockholm Contr. Geol.* 2(5):73–102.

Kurtén, B. 1960. A skull of the Grizzly Bear (*Ursus arctos* L.) from Pit 10, Rancho La Brea. Los Angeles County Mus. *Contr. Sci.* 39:1–6.

Kurtén, B. 1964. The Carnivora of the Palestine Caves. *Acta Zool. Fennica* 107.

Merriam, J. C. and C. Stock, 1925. Relationships and structure of the Short-Faced Bear, *Arctotherium,* from the Pleistocene of California. Carnegie Inst. Washington Publ. 347(1):1–35.

Nordmann, V. and M. Degerbol, 1930. En fossil Kaebe av Isbjøn (*Ursus maritimus* L.) fra Danmark. Vidensk. *Med. Dansk Naturhist. Foren.* 88:273–286.

Ognev, S. I. 1935. *Mammals of U.S.S.R. and Adjacent Countries*. (In Russian.)

Hecht, Max K. 1963. The role of natural selection and evolutionary rates in the evolution of higher levels of organization. Proc. XVI Internat. Congr. Zoology, Washington, 20–27 August 1963 3:305–308.

Thenius, E. 1953. Zur Analyse des Gebisses des Eisbären, *Ursus (Thalarctos) maritimus* Phipps, 1774. Saugetierkundl. *Mitteil*. (1953 1):1–7.

Tscherski, J. D. Wissenschaftliche Resultate der von der Kaiserlichen Akademie der Wissenschaften zur Erforschung des Janalandes und der Neusibirischen Inseln in den Jahren 1885 und 1886 ausgesandten Expedition. Abäth. 4. Beschreibung d. Samml. posttertiärer Säugethiere. *Mem. Acad. Imp. Sci*. St. Petersburg, ser. 7(1):40.

Zapfe, H. 1946. Die altplistozänen Bären von Hundsheim in Niederösterreich. *Jahrb. Geol. Bundesanst*. 1946(3–4):95–164.

Zeuner, F. E. 1959. *The Pleistocene Period*. London: Hutchinson.

Zimmermann, 1945. Mittheilungen an den Geheimenrath von Leonhard gerichtet (Briefwechsel). *N. Jahrb. Geol*. etc., for 1945, pp. 73–74.

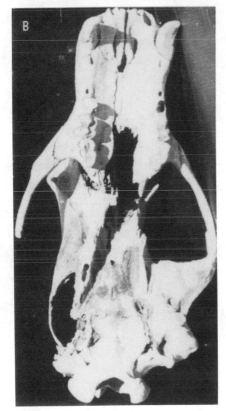

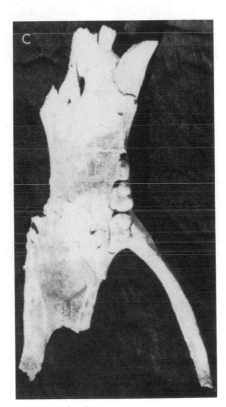

Plate 4.1. Subfossil finds of the Polar Bear.

A. *Ursus maritmus,* subfossil, N.H.M.G. 1925–4354, skull from Kärraberg, side view.
B. Same specimen as A, ventral view.
C. *Ursus maritimus,* subfossil, N.H.M.G. 1919–3356, skull fragment from Hisingen, palatal view.

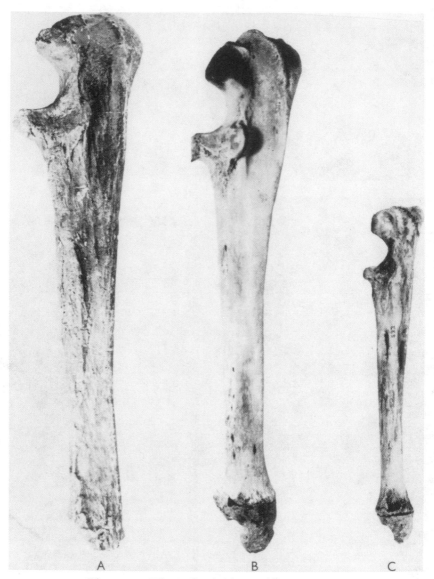

Plate 4.2.   Ulnae of ursine bears. All to same scale.

A. B.M. No. 24361, type of *Ursus maritimus tyrannus,* new subspecies. Pleistocene, Kew Bridge, London.
B. U.Z.I. No. 3, *U. maritimus maritimus,* recent.
C. N.R.M. No. 233, *U. arctos arctos,* recent, Sweden.

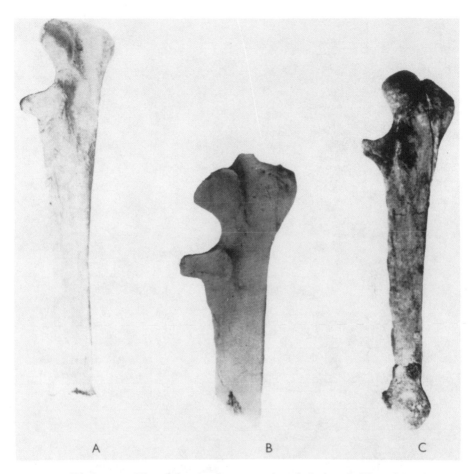

Plate 4.3.   Ulnae of ursine bears, continued. Scale as in Plate 2.

A. B.M., no number, *U. spelaeus,* Pleistocene, probably from Gailenreuth.
B. B.M. No. 48832, *arctos priscus,* Pleistocene, Brixham, proxima lfragment.
C. B.M. No. 10500, *U. deningeri* (or *U. savini*), Pleistocene, Bacton Forest Bed.

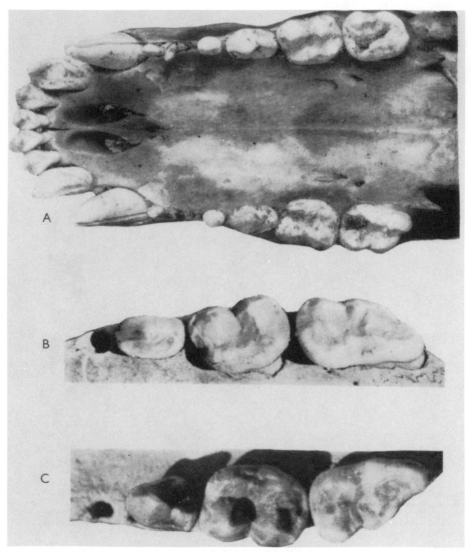

Plate 4.4. Upper dentitions of Polar Bear.

A. C.N. No. 1859, recent, Greenland. Juvenile specimen with partially (right) or completely (left) reduced talon in M^2. Not to same scale as B and C.
B. Left cheek teeth of subfossil specimen from Hisingen (Plate 4.1 C). P^4 without protocone.
C. Right cheek teeth of subfossil specimen from Kärraberg (Plate 4.1 A and B). P^4 with well-developed protocone.

FIVE

Return of a Lost Structure in the Evolution of the Felid Dentition

INTRODUCTION

THE LOWER CARNASSIAL in the living Felidae is generally formed by the trigonid only, bearing two cusps, the paraconid and protoconid. The vestige of the talonid may be represented by a small basal bulge at the hind edge of the tooth. In some forms, however, especially in the subgenus *Felis (Lynx)*, the talonid may be somewhat larger, and in addition there may be a cusp or style along the hind edge of the protoconid; it is usually coalescent with the protoconid and does not have a separate point. This cusp is generally called the metaconid, though the homology as such is not always accepted. A somewhat similar character is seen in many fossil Felidae.

The metaconid-talonid complex may be termed a postcarnassial dental element in the sense of Crusafont and Truyols (1958), from its position behind the carnassial shear and within the field of molarization (Butler 1939). Its presence is usually considered a primitive character, and may probably be so in the majority of cases. It was surprising, however, to find that the condition in the living *Felis lynx* seems to be derived from one in which the complex is absent or almost absent (Kurtén 1957). Even more astonishing, this seems to be coupled with the re-appearance of M_2, a structure unknown in Felidae since the Miocene. All of this, of course, is completely at variance with one of the most cherished principles of evolutionary paleontology, namely Dollo's Law.

The Pleistocene and recent lynx material mentioned here has been studied in recent years in the collections in London, Paris, Basel, Hei-

Reprinted, with permission, from *Commentationes Biologicae* (1963) 26 (4):1–12.

delberg, Lund, Uppsala, Stockholm, and Helsingfors. In addition, other fossil felids have been studied in the American Museum of Natural History, New York, and the Yale Peabody Museum, New Haven. Grants-in-aid have been received from the Rockefeller Foundation, Statens Naturveten-skapliga Kommission, and the University of Helsinki/Helsingfors. I wish to express my sincere gratitude to all the institutions and persons concerned.

EVOLUTION OF EUROPEAN LYNX

Early Villafranchian

The earliest Villafranchian deposits in Europe are represented by the Etouaires horizon of the Montagne de Perrier. They carry a lynx-like cat known as *Felis issiodorensis* Croizet & Jobert. The species survived to the end of the Villafranchian (upper Val d'Arno). It is characterized by numerous lynx-like characters in the skull and dentition, but the post-cranial skeleton is of a generalized type with a relatively long body and short legs. The head is relatively larger than in living lynxes.

In this early form the lower carnassial has an almost straight posterior edge (fig. 5.1). There is no trace of a metaconid, and the talonid is repre-sented only by a very slight bulge, without structural detail, as in the usual recent cat type. This condition was found in all of the five specimens examined (representing five individuals), and may thus be inferred to represent the modal, if not the only, type present in the population.

Middle and late Villafranchian

The following stages in the evolution of *Felis issiodorensis* are represented by the material from the upper levels of Perrier (Pardines; early middle Villafranchian), Saint-Vallier (Viret 1954; later middle Villafranchian), and Val d'Arno (Fabrini 1896; late Villafranchian). Pardines has yielded a fauna with some typical Villafranchian immigrants (*Equus bressanus, Leptobos etruscus,* and others) but still without *Elephas;* by Saint-Vallier times, how-ever, true elephants are present.

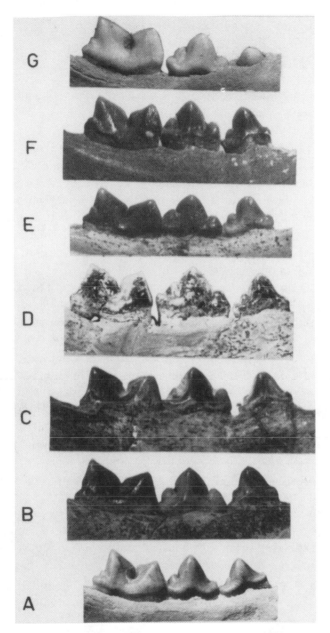

Fig. 5.1. Lower premolars and carnassials of fossil Felidae, representing the left jaw, in lingual view (some figures reversed from right jaw). All specimens in the Natural History Museum, Basle. A, No. 16178, *Pseudaelurus lorteti* Gaillard, Miocene, Steinheim; to show lynx-like M₁ in early felids. B, No. Prr. 204, and C, No. Prr. 200, both from Etouaires, showing absence of metaconid and rudimentary talonid in M₁ of earliest *Felis issiodorensis*. D, No. L.P. 150, *F. issiodorensis*, Pardines, and C, No. V.A. 1164, *F. issiodorensis*, Val d'Arno, with metaconid and talonid present. F, No. Sl. 15, *F.* sp., Blanzac, middle Pleistocene. G, No. B 65 V, *F. lynx*, Veyrier near Geneva, late Pleistocene (Magdalenian), juvenile specimen with emerging premolars. All 1½ times natural size.

Out of six carnassials from the upper levels of Perrier, one is of about the same type as those from Etouaires. Four have a more distinct talonid bulge, and one, finally, has not only the talonid but also a distinct metaconid.

In the sample from Saint-Vallier, the same types are represented, according to the descriptions and figures in Viret (1954). One specimen out of five mentioned has a metaconid. The relative frequencies of the three types appear to be about the same as in the slightly earlier Perrier sample.

In the late Villafranchian lynx from Val d'Arno, described and figured by Fabrini (1896), the talonid appears always to be fairly well developed; in other words the Etouaires type has vanished. The metaconid is present in one and absent in four of the specimens figured.

Two recent species exhibit a similar type of variation. One of them is *Felis pardina* Temminck, the small lynx from the Iberian Peninsula. (Some authors consider *Felis pardina* as a subspecies of *F. lynx,* but in my opinion the evidence for specific separation is valid. In addition to the morphological differentiation there is evidence that the ranges of the species overlapped in the Pleistocene without resultant hybridization.) In 11 specimens of this form seen by me (10 in the British Museum, one in Riksmuseet, Stockholm) the talonid is uniformly well developed; but seven lack the metaconid completely, three have a very weakly developed metaconid, and one has a somewhat more developed metaconid. The other species is the desert lynx, *Felis caracal* Schreber, of which 23 specimens were examined (in the British Museum). Sixteen have no trace of the metaconid; one shows a slight trace of the cusp, and two more specimens have it very poorly developed; in two other specimens, finally, it is almost as large as in *Felis lynx.* All this shows that an intraspecific variation of this type is nothing unusual, and that Viret (1954) was fully justified in uniting the Villafranchian forms in a single species, though they have been split up into different species by earlier authors.

Middle Pleistocene

I have seen only two lynx specimens showing the character of M_1 and dating from the middle Pleistocene. They are from Blanzac-Solilhac (in the Basel Museum) and Mauer (Heidelberg collection), and in both there is a small talonid and a small metaconid. They resemble *Felis pardina* rather than *Felis lynx,* and it is possible that they may be regarded as representatives of an early, large form of the pardel lynx. As late as the last glaciation, Spanish *Felis pardina* reached almost the same size as the Northern lynx.

Late Pleistocene and Recent

In the living lynxes quite varied conditions are found in different species. In some forms, such as *Felis rufus* Schreber, the metaconid is rarely seen, and the modal condition seems to resemble that found in the early Villafranchian *Felis issiodorensis. Felis canadensis* Desmarest has a talonid in M_1, but the metaconid appears usually to be absent. The situation in *Felis caracal* and *Felis pardina,* described above, resembles that in the middle and late Villafranchian *Felis issiodorensis.* In *Felis lynx* L., finally, or at any rate in the European material of this species known to me (about 60 skulls examined), the metaconid-talonid complex is a fixed character, present in all the carnassials, though the size of the two elements is somewhat variable. In late Pleistocene specimens of *Felis lynx, Felis pardina,* and *Felis rufus* seen by me, the characters are quite similar to those in the present-day forms.

Table 5.1 shows the relative frequencies of the various types. The fossil series shows a gradual shift of the mode from an early condition without

Table 5.1.
Relative frequencies of three morphotypes of M_1 in fossil and Recent samples of *Felis (Lynx)*.

	N[a]	A (%)	B (%)	C (%)
Felis lynx, Recent	60	0	0	100
Felis caracal, Recent	23	0	70	30
Felis pardina, Recent	11	0	64	36
Felis sp., middle Pleistocene	2	0	0	100
Felis issiodorensis, Val d'Arno	5	0	80	20
Felis issiodorensis, Pardines	6	17	66	17
Felis issiodorensis, Etouaires	5	100	0	0

a N, number of specimens; A, no talonid, no metaconid; B, talonid present, metaconid absent; C, talonid and metaconid both present.

talonid and metaconid to the final *Felis lynx* stage in which both elements are invariably present. The other living species may be thought of as having branched from this line at various points.

FOSSIL LYNXES IN CHINA

Outside Europe the most important evidence on lynx evolution is found in China. The material has been surveyed by Teilhard (1945). A Villafranchian form is described as *Lynx shansius* Teilhard; the material comes from Loc. 12 at Choukoutien, the Sanmenian of Nihowan, and Loc. 18 near Peking. The sequence is comparable to the middle and late Villafranchian in Europe. Teilhard also mentions a large material from the Villafranchian of Yushe. In all the lower carnassials (total number about 9?) of this form, the metaconid is stated to be present; an accompanying figure shows the metaconid-talonid complex to be fairly similar to that in the Val d'Arno lynx. The condition is similar in the jaw from Nihowan in the Paris collection. It is doubtful whether this form is specifically distinct from *Felis issiodorensis*.

Teilhard also mentions a lot of eight mandibles from Tsao Chuang in Shansi, in the Frick Collection. All the carnassials in this material are stated to lack the metaconid. Possibly this sample may represent the Chinese wing of the early Villafranchian population from the basal Villafranchian of Etouaires. The size of this form is the same as that of *Felis shansius*.

Another group of forms, *Felis peii* Teilhard from the Villafranchian and *Felis teilhardi* Pei from the middle Pleistocene, generally resembles the lynxes but differs by the presence of P^2. The lower carnassial of the earlier form is not definitely known (unless the jaw described as ? *Lynx* sp. 1 belongs here), but the material of *Felis teilhardi* from Choukoutien varies in the same manner as the European lynxes: one specimen has neither metaconid nor talonid, two have a talonid, and one has the entire complex.

The true *Felis lynx* occurs in the late Pleistocene Upper Cave at Choukoutien.

The Chinese sequence appears to be closely comparable to that in Europe, as far as the available data show. Data on lynx evolution in North America are very incomplete, but *Felis rexroadensis* Stephens (1959) may be close to *Felis issiodorensis*. In the American form P^2 was still present, but of minute size, and tended to be lost in life. Unfortunately the lower carnassial is unknown.

INTERPRETATION

There is a graded series of transitions, both in China and in Europe, with overlapping variation in successive populations. This must be regarded as strong evidence that the sequence is a phyletic one and hence that lynxes with metaconid and talonid in M$_1$ evolved out of ancestral forms lacking both. The main objection to this interpretation would be the fact that the character thus evolved shows a striking resemblance to those found in geologically earlier felids, probably including the direct ancestors of *Felis issiodorensis,* though they are not at present known or at least not identified. A good example is the Miocene genus *Pseudaelurus* (including the Pliocene *Metailurus*), shown in fig. 5.1 together with some representatives of the lynx line. This would then be an example of a structure totally lost and then regained in similar form—which is something that simply "cannot" happen according to Dollo's Law.

Of course it would be possible to interpret the facts in another more "orthodox" way, as due to repeated migrations of successively more "primitive" forms from an unknown center of distribution. Studied in detail, however, this proposition becomes so complicated as to border on the miraculous. The main argument for the interpretation suggested here is, however, the reappearance of another dental structure of vastly greater geological age, in the living *Felis lynx*. This is the second lower molar, which was lost in the Miocene in all known Felidae.

PRESENCE OF M$_2$ IN THE NORTHERN LYNX

The second lower molar occurs as an extremely rare individual abnormity in recent cats. Teilhard mentions a Chinese wildcat ("Chat sauvage," species not identified) in the "Musée Heudes," Shanghai, and a Chinese leopard cited by Gray. In these cases the molar is present on one side only. The presence of M$_2$ in *Felis lynx* cannot be regarded as abnormal; its frequency is high enough to show that it belongs to the normal variation pattern of the species, and it may be present on both sides. Out of 22 skulls from Finland, three have M_2; and out of 31 specimens from Sweden, two have M_2. The difference between the relative frequencies in the two samples is unimportant, and the average frequency may be set as five out of 53, or about 10 percent (more exactly 9.4 ± 3.4).

One of the Finnish specimens has M_2 on both sides. In this case the second molars are small round pegs, peculiarly situated on the lingual side of the carnassials. In a second individual the tooth is normally placed behind the carnassial, and similarly a small rounded structure. The third Finnish jaw (fig. 5.2) has a larger M_2, 5.0 mm in length and 3.3 in width, of distinctly molariform shape. In occlusal view it has an ovoid outline, the posterior part being broader; it is orientated obliquely to the long axis of the ramus, the front end being lingual to the talonid of M_1. In side view the molar resembles the M_2 of some canids to some extent, but it is also reminiscent of the lynx M_1. It has a broad blade, consisting of a low

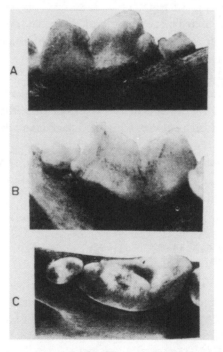

Fig. 5.2. *Felis lynx,* recent, Finland, right M_1—M_2, A internal, B external, C occlusal view. Museum of Zoology, University of Helsingfors. Twice natural size.

paraconid and a slightly higher protoconid. On the lingual slope of the latter there is a very low, hardly noticeable swelling, which would be in the normal (non-felid) position of the metaconid. On the other hand, there is a minute ridge which, as in M_1, borders the posterior edge of the protoconid to about two-thirds of its height. This ridge coalesces with the talonid, which is simply a slight posterior bulge, and the structure is thus reminiscent of the metaconid-talonid complex in M_1.

None of the Swedish specimens has such a large M_2. In one the tooth is present on both sides and has a length of 2.7 and a width of 2.8 mm. The other specimen has only the left M_2, and the dimensions are 2.5 by 2.5 mm. These teeth are thus small round pegs, as in two of the Finnish individuals.

Our knowledge of the early Tertiary Felidae is still too limited to permit reconstruction of the ancestry of the living cats. They must, however, have descended from forms with a molariform M_2. This structure was lost in all known Felidae in the Miocene. It has not been observed in any fossil lynx. Its presence in an appreciable proportion of the present-day Northern Lynxes may thus be regarded as evidence of an evolutionary reversal of the same type as the return of the metaconid-talonid complex in M_1.

It seems most probable that the appearance of M_2 is actually coupled with the development of the metaconid-talonid complex. Butler (1937, 1939) studied the processes of molarization and caninization and suggested a field effect, probably related to the morphogenetic fields regulating the growth processes. To follow this line of thought, the development of postcarnassial or post-shear elements in the lynx lineage may have been brought about by a reactivation of the molarization field, resulting in a size increase of this part of the dentition. The activation might have proceeded far enough to bring M_2 above the size of the realization threshold (see Kurtén 1953) in some individuals. This, of course, presupposes that M_2 was never genotypically lost.

The steady continuation of the trend in one lineage throughout the Pleistocene suggests that it has some adaptive significance in the ecological situation of the lynx. It does have the effect of adding a chopping mechanism to the dentition, but in what way this may be useful to the species is unknown. However this may be, the trend is a most striking and surprising exception to the dictum that lost structures never reappear in their original form.

REFERENCES

Butler, P.M., 1937. Studies of the mammalian dentition. I. The teeth of *Centetes ecaudatus* and its allies. *Proc. Zool. Soc. London*, Ser. B., vol. 107.

——1939. Studies of the mammalian dentition. Differentiation of the post-canine dentition. *Ibid.*, ser. B, vol. 109.

Crusafont Pairó, M., & J. Truyols Santonja, 1957. Estudios masterométricos en la evolución de los fisipedos. *Bol. Inst. Geol. Min. España*, vol. 68.

Fabrini, E., 1896. La lince del pliocene italiano. *Palaeontogr. Ital.*, vol. 2.

Kurtén, B., 1953. On the variation and population dynamics of fossil and recent mammal populations. *Acta Zool. Fennica*, (no. 76):p1–122.

——1957. A note on the systematic and evolutionary relationships of *Felis teilhardi Pei Vertebr. Palasiatica*, vol. 1, (2):p123–128.

Stephens, J. J., 1959. A new Pliocene Cat from Kansas. *Papers Michigan Acad. Sci.*, 44:p41–46.

Teilhard de Chardin, P., 1945. Les formes fossiles. In Chardin & P. Leroy, *Les Félidés de Chine* pp. 5–35, Peking 1945.

Viret, Jean, 1954. Le loess à bancs durcis de Saint-Vallier (Drôme) et sa faune de mammifères villafranchiens. *Nouv. Arch. Mus. Hist. Nat. Lyon*, 4:p1–200.

III
PALEOBIOGEOGRAPHY

SIX

Mammal Migrations, Cenozoic Stratigraphy, and the Age of Peking Man and the Australopithecines

INTRODUCTION

STRATIGRAPHIC ZONATION of terrestrial Cenozoic deposits is largely based on the changes in their mammalian faunas. The changes result from (a) extinction, (b) evolution, and (c) migration.

The weight given to each of these factors, in an attempt to determine the temporal relationships of two faunas from different localities, differs from case to case, and this is proper. But in general they can be ranked according to their importance in correlation, and then the sequence as given is one of increasing importance.

The stratigraphic significance of a faunal change is relative to its temporal and distributional rate. In the zonation of a single stratigraphic column, (a) and (c) denote points in time, whereas (b) is a gradual process. In the comparison between two stratigraphic columns, (a) loses much of its significance, for a phylum may linger on in one area for a long time after it has been exterminated in another. Evolution (b) is generally held to be more important; relative levels of evolution in all the phyla constituting the two faunas give fair indication of their temporal rela-

Excerpted, with permission, from *Journal of Paleontology* (1957) 31:215–227.

tionship. In some instances the evidence from (b) has even been considered to overrule that from (c), migration. More often, however, the immigration of a new phylum at two localities is considered to be, geologically speaking, simultaneous, as long as the geographic distance between them is reasonably short. The validity of long-range correlation on this basis has been in doubt, because the time-rate of mammalian migrations has been obscure. Available data on Recent migrations and species ranges shed important light on this problem.

RATES OF MIGRATION

The migration of a species takes the form of population spread. It thus differs from such phenomena as the annual migration of birds (and some mammals and other animals); but it should be noted that some borderline cases are known, where more or less concerted mass movements do lead to the acquisition of new territories. A famous and long misinterpreted instance is that of the lemmings of northern Europe. The popular idea is that the lemmings, in years of proliferation, march to their death, the whole "army" ultimately perishing, whereas the population is carried on by the few stationary individuals. Recent studies (Kalela 1949) have shown that this picture is largely incorrect. The lemming migrations generally lead to settlement in new areas, whereas the original territory is left with a much attenuated population. In this way the center of the population is rotated, often returning to the original area after a series of shifts.

But this is exceptional, and population spread is the common mode of migration. It results from the random movements of the individuals. The basic unit is the average distance between the birthplace of an individual and that of its parent, and the standard deviation of this distance. In principle the movement may be in any direction; where it is directed outward from the boundary of the species range, the range is extended. It is a matter of observation that advances of the range are initiated by vagrant individuals moving outside of the settled area of the species.

Such a range extension is commonly checked by ecological barriers—physical, physiological, and others. Actual migration takes place with the removal of the barrier. This may happen in two different ways. In the first place, the barrier may be slowly retreating (say, a shift of a climatic zone, with consequent shifts of vegetational belts), and the spread of the population will keep pace with the retreat of the barrier. In the second place, there may be an ecological threshold; once it is crossed or vanishes, a large new area is open to colonization. Then an intermediate situation may be visualized: the barrier may retreat so rapidly that the species fails to catch up with it.

In the initial case, the rate of migration will be determined by the rate of change in the position of the barrier. In the two latter instances, the rate of migration depends entirely on the spreading capacity of the population.

During the Pleistocene in particular, the former mode was very common. With the advance of glaciers and the concomitant retreat of temperate biotopes, animals of arctic type spread southward. With the shrinking of ice sheets and the northward retreat of arctic biotopes, populations spread northward.

Most instances of such population spread, mainly dominated by gradual shifts of ecologic zones, probably also contain, as details in the main picture, minor instances of "threshold-effect" migration (a Recent case will be discussed below). But the most typical examples of this kind of migration occur when an obstacle is forced, bridged, or disappears. The spread will then be especially rapid if the conditions in this new area are similar to those in the original territory, because then no marked shift in the adaptation of the population is required.

What is the possible rate of such a colonization, and which mammals are especially likely to spread rapidly, and thus be particularly useful in correlation?

The study of individual migratory ranges is unfortunately still in its beginning, and most results so far have been obtained by the banding of birds and small mammals—which are generally rare as fossils and often taxonomically difficult. A good fossil record and sound taxonomy are indispensable for good results. However, there are a few well-analyzed instances of present-day migration, concerning mammals of the type that are common as fossils; these have been studied particularly by Kalela (1940, 1948), whose results will be briefly summarized here.

From 1880 to 1940, the polecat (*Putorius putorius*) advanced in Finland from the Carelian Isthmus westward and northwestward to the coast of the Bothnian Gulf, as far north as central Ostrobothnia. The longest distance covered was about 450 km, and the rate of migration thus averaged some 7.5 km annually, or 750 km in a century.

As shown by Kalela, and as indicated by many other range changes during the same period (the roe deer, to be discussed below, and many bird populations), the advance was mainly a response to the climatic amelioration during this time. It should however be noted that the main direction of the advance was from east to west, not south to north; thus it seems probable that, in this special case, there may have been a threshold effect. It may perhaps be suggested that the initiation of a warmer climate permitted the population to spread from the Carelian Isthmus, whereby the road westward was opened; thus the main part of the migration later on may have been unchecked. The population was channeled in a western and northwestern direction because further northward spread was checked by a climatic barrier. The final situation, with a northern species boundary running from northwest to southeast, is simulated by many other species of animals and plants in Finland, and evidently reflects a climatic zonation.

The roe deer *(Capreolus capreolus)* has an even more spectacular migration history in about the same time. In 1850 there existed only a very small population in a limited part of Scania in southernmost Sweden (slightly south of the 56th parallel). During the next 95 years, the species spread into great parts of Sweden and Norway, reaching the 64th parallel and even farther north. The distance covered was 900 km or more, so that an average rate of about 1000 km in 100 years is indicated. It should also be noted that the rate was low initially, probably because population pressure was low in the small original population, and the spread was accelerated from about 1880, when the pressure presumably mounted with population size.

The spread of the roe deer is thought mainly to depend on climatic change and the extinction of dangerous predators, particularly the wolf. Direct human activity appears to have played an insignificant role.

The seeming difference in the rate of migration between polecat and roe deer may suggest that this rate of positively correlated with size. Presumably such a correlation would not be absolute, but it is not unlikely to hold good in general. Probably, fleetness of foot and rates of reproduction are also important.

There is no reason to suppose that the polecat or the roe deer possess unusual spreading capacity. The polecat is a relatively small carnivore with a rather narrow specialization; the roe deer is highly dependent on mild winters for its subsistence. The figures would thus seem to be representative of common migration rates in mammal species. Larger and more ubiquitous mammals may be able to spread more rapidly. An unchecked spread of some 1000 km in a century would seem a moderate estimate for most larger mammals.

This rate would mean that, under favorable circumstances, a mammalian species may migrate from the Bering Straits to Western Europe in about 1000 years, a time unit which is negligible in the Cenozoic geological history. And even if the rate of 1000 km per century were an overestimate, and the true rate for a species were, perhaps, only one-half or one-quarter of that, the difference would have no significance whatever in the geological time scale. Unchecked spread denotes a point in geological time. This spreading capacity appears to have been vastly underestimated in much paleontological thought.

This, it should be remembered, holds for unchecked spread only. Clearly the migration of many mammals has been much slower, and many evolving phyla have failed to colonize more than a limited area. The adaptation of most species is so narrow that optimal conditions obtain in limited areas only (and even if similar zones are available in other areas, intervening barriers may preclude migration); hence, for a colonization of new areas, evolutionary change is necessary. If the environmental gradient is steep, such a change becomes unlikely. Genetic adjustment to the different environment will often be swamped by the mother population, and only the accidental isolation of a small subpopulation is likely to lead to adaptation.

This is probably a factor which intervenes, to some extent, in all large-scale migrations into great new areas. The necessary change may sometimes be effected by means of very small evolutionary steps (a single mutation in the simplest case) and the evolution will then be on the deme or subspecies level; if so, the migration will hardly be much repressed. But the adjustment may also touch many genes, so that the evolution will be on the species level or higher; the rate of migration would then be arrested in a roughly correlative way. Thus it is clear that the taxonomic relationship between immigrants at different sites gives important clues to the rate of migration and hence to the precision of the correlation; and the taxonomic treatment assumes the highest importance.

PRESENT-DAY SPECIES RANGES

The immigration of a single species at two localities is an event of great stratigraphic value. What is the likelihood of such an event occurring at two localities far apart, and which kinds of mammals may be especially useful in such long-range correlation?

To obtain an answer, we must first study the species ranges of Recent mammals. A study was made of some mammalian families and one subfamily typical of the modern fauna—the successful mammals of the present day. Among carnivores, the Canidae, Ursidae, and Felidae were selected, and among herbivores, the Bovidae and the Cervinae. The interest was focused on the larger carnivores and ungulates for several reasons. In the first place, the larger mammals are not only most common and best understood in fossil faunas, but they also in general have the widest species ranges and may be especially useful in long-distance correlation. Secondly, in the Recent fauna, the taxonomy of the selected groups is fairly well worked out, and the species ranges are relatively well known. Obviously a study of rodent ranges would be of great interest, and the results would be valuable for generalizations on account of the great number of species; but rodent taxonomy is not satisfactory, and data on distribution are vague. The same holds for many other small mammals.

The ranges of distribution were computed from Ellerman and Morrison-Scott (1951); for the Ursidae, Erdbrink (1953) was also used. The figures thus obtained will have some tendency to bias, giving a somewhat greater range than the actual; the distributions of most species are patchy, but more precise data can be obtained only for very few species, and the treatment should be consistent. From a paleontological point of view, in fact, the figures used here are more enlightening than very accurate estimates giving only the sum of the population "islands" within the total range.

The species which occur in the Palearctic and Indian regions were used, this being the scope of Ellerman and Morrison-Scott's compilation, but for the species which occur also outside of these regions, the total ranges were taken. The result is probably a slight increase in group means, because endemic forms from other regions, presumably often with fairly small ranges, are left out. As the region covered is the largest land mass of the world, however, this effect can hardly be very important, and the bias is likely to be consistent.

The results are tabulated in table 6.1. The species ranges are given in million square miles, in the form of frequency distributions, and the statistical parameters (arithmetic mean range and standard deviation) are entered for each distribution.

The mean ranges for species of the Canidae, Felidae, and Ursidae are all approximately the same (the slight deviation for the bears lacks significance), and the similarity of the values seems to indicate that the carnivorous families tend to approach a certain general pattern of distribution. The combined figures for the three families give an estimate of this pattern.

The bovid pattern is quite different. The mean is here much lower, little more than one-fourth of the carnivore mean. The cervine mean takes an intermediate position, being roughly twice the bovid mean, and half of the carnivore mean.

Like the means, the maximum ranges of carnivorous species exceed those of ungulates. The most wide-ranging carnivore *(Vulpes vulpes)* cov-

Table 6.1.
Distribution of Species Ranges (in Million Square Miles) for Four Families and One Subfamily of Mammals
(Only species present in the Palearctic and Indian regions included)

	Number of species					
Range	Canidae	Felidae	Ursidae	Canidae + Felidae + Ursidae	Bovidae	Cervinae
0– 1	3	0	0	3	20	8
1– 2	0	4	2	6	8	2
2– 3	1	1	0	2	4	0
3– 4	2	2	1	5	3	1
4– 5	2	0	1	3	2	1
5– 6	0	2	0	2	2	0
6– 7	2	1	0	3	1	0
7– 8	0	2	0	2	0	0
8– 9	0	1	0	1	2	0
9–10	0	0	0	0	0	0
10–15	1	4	0	5	0	2
15–20	1	1	1	3	0	1
20–25	0	0	0	0	0	0
25–30	1	0	0	1	0	0
Total	13	18	5	36	42	15
Mean range	6.8±2.1	6.8±1.1	5.9±3.0	6.7±1.0	1.8±0.3	3.7±1.3
Standard deviation	7.7±1.5	4.6±0.7	6.7±2.1	6.2±0.7	2.2±0.2	5.1±0.9

ers about three times the areas of the most widely distributed bovids *(Gazella dorcas, Capra ibex)* and about twice that of the three widespread cervines *Capreolus capreolus, Cervus elaphus,* and *Alces alces.*

This suggests that the larger carnivores will be especially valuable in long-distance correlation. It also shows that the not uncommon habit of placing populations from localities far apart in different species, simply because of the great geographic distance, is wholly unwarranted, in particular when the Carnivora are concerned.

The shape of the distributions in table 6.1 is of some interest. They are more or less asymmetric and generally have a mode lower than the mean, but some appear to be bimodal or multimodal. The standard deviation is about equal to the mean. This is a character of the Poisson distribution, and a few of these distributions approach the Poisson type.

The J-pattern is most clearly seen in the bovid distribution. Here the 0–1 million square miles group is also the modal one, and the frequencies fall off regularly toward the higher values, except for a small secondary concentration in the 8–9 million square miles class. This impression still holds if the initial class is subdivided:

Species ranges	0.0–0.2	0.2–0.4	0.4–0.6	0.6–0.8	0.8–1.0
No. of bovid species	10	3	3	2	2

With still finer subdivision, of course, a decrease of frequencies to the left of the mode can be demonstrated (though the actual data are not accurate enough to permit a detailed study), for a species range cannot be infinitesimal.

The distributions for other groups depart more from the J-Pattern and may suggest a combination of several different distributions. Both the combined carnivore distribution and the cervine distribution suggest three distinct modes—one with small species ranges (0–2 million square miles), another with medium ranges (3–8 million square miles) and a third with very great ranges (10 million square miles and more). This gives a division into roughly 25 percent small-range species, 50 percent medium-range species, and 25 percent wide-range species among the carnivores. For the cervines, the corresponding percentages would be on the order of 65, 15, and 20 respectively; here the small-range species predominate. The bovid distribution may suggest a combination of a greatly preponderant distribution of small-range species (the range of which is extended by the great number of observations) and a small group of medium-range species; the percentages would be 95, 5, and 0.

What is the cause of the greater average and maximal ranges of carnivorous species?

The adaptation of most carnivores is broader than that of herbivores in general. A carnivorous mammal may live on many different kinds of prey, whereas most herbivores are specially adapted to certain vegetational zones.

The historical processes conducive to greater species ranges in carnivores may be visualized as follows. An ungulate species may spread into an area containing, say, two main biotopes. Optimal adaptation to each of these necessitates genetic differentiation. As long as the two populations interbreed in the marginal zone, however, such differentiation cannot become complete; there will be some gene flow across this zone. Selection acts against the gene flow, and the tendency is to build up a genetic barrier, leading ultimately to speciation.

The carnivore that immigrates into the same area will need less genetic differentiation in order to cope with the different environments, because its adaptation is more ubiquitous; consequently, the tendency to form a genetic barrier is weaker, and speciation is slow.

The result will be much the same if an otherwise uniform region is divided by a zone with different ecological conditions—say, a steppe crossed by a forest belt. The more narrowly adapted ungulate populations will be divided by this barrier, which permits only restricted gene flow or none at all; whereas a carnivore may retain genetic contact across it. Here, too, the ungulate will speciate more rapidly than the carnivore.

Another important factor is population size. In a given area, the herbivorous mammals always outweigh the carnivorous. An ungulate population is much denser than that of a carnivore of comparable size. This leads in the first place to a much greater concentration of ungulate species with very small ranges. Most of these are relict species, the range of which has shrunk to its present size. When the number of individuals in a population is constantly held below a certain minimum, extinction is inevitable. For the ungulate population, with its high density, this limit size corresponds to a smaller range than for the carnivore. Moreover, the limit number will be lower because of greater facility for pairing in the dense population.

The lower population density of carnivores probably induces greater individual movement and more rapid gene flow in the population, and thus counters speciation.

Another concomitant of the lower population density of carnivorous species may also be of stratigraphic importance. Their rates of evolution may be somewhat higher than those of the larger ungulate populations. There is actually some indication that this may be the case. Rates of evolution in the Equidae (Simpson 1953) and Oreodontidae (Bader 1955) are lower than in the Ursidae (Kurtén 1955). For a valid generalization, however, more data are needed.

SPECIES RANGES IN THE LATE CENOZOIC

Were the distributions of species ranges for the dominant mammals of earlier epochs of the same general type as those found in present-day faunas?

No Cenozoic fauna is well enough known from all parts of the Palearctic and Indian regions to permit computation of total species ranges, and a different approach is necessary. The method is now to select some well-documented faunas and study their representatives at localities relatively far apart, ascertain the number of species common to both areas, and compute their relative frequency. Comparison with Recent conditions will indicate to what extent the distributions conserve their pattern through time.

I have selected the European and Chinese faunas of the Pontian (lower Pliocene), of the Cromerian (Gunz-Mindel integlacial), and of the latest Pleistocene. The groups studied are the Canidae, Ursidae, Felidae, and Hyaenidae (the last-mentioned group was not considered in the study of the Recent distributions because it has so few surviving members). The comparisons are given in tables 6.2–6.4; the appended notes give references and some brief explanations of the taxonomic treatment where it differs from previous opinions. Some species of uncertain status have not been included.

The number of species known in Europe and China, respectively, at different times is given under N_1 and N_2 in table 6.5, and the number of species common to both areas under C. The Chinese forms bracketed in table 6.2 are not included in the tabulation, because of their uncertain

Table 6.2.
Carnivores (Canidae, Ursidae, Hyaenidae, and Felidae) of the European and Chinese Lower Pliocene

	Europe	China
Canidae	*Amphicyon pyrenaicus*	(*Amphicyon* sp.)
	Simocyon primigenius	Simocyon primigenius
	Metarctos diaphorus	
Ursidae	*Indarctos ponticus ponticus*	*Indarctos ponticus lagrelii*
	Indarctos atticus	(*Agriotherium* sp.)
	Ursavus depereti	
Hyaenidae	*Crocuta eximia eximia*	*Crocuta eximia variabilis*
		Ictitherium gaudryi
	Ictitherium robustum	
	Ictitherium hipparionum hipparionum	*Ictitherium hipparionum wongii*
	Hyaenictis graeca	*Ictitherium hyaenoides*
	Lycyaena chaeretis	
Felidae	*Machairodus aphanistus aphanistus*	*Machairodus aphanistus palanderi*
		Machairodus tingii
	Epimachairodus taracliensis	
	Paramachaerodus orientalis orientalis	*Paramachaerodus orientalis maximiliani*
		?*Megantereon* sp.
	Metailurus parvulus parvulus	*Metailurus parvulus* minor
		metailurus major
	Felis attica	(*Felis* sp.)
	Felis neas	

Notes: *Indarctos lagrelii* is clearly close enough to *I. ponticus* to be considered the same species; the distinction was mostly based on the geographic distance. The same holds for *Crocuta variabilis* (see also Orlov, 1941), *Ictitherium wongii* (see Kurtén, 1954b), and *Machairodus palanderi* and *Paramachaerodus maximiliani* (see Pilgrim, 1931), as compared with their European allies. The relationships between "*Felis leiodon*," "*Machairodus*" *parvulus*, and *Metailurus minor* have been cleared up by Thenius (1951). *Indarctos sinensis* is probably a male of the species of which the type of *I. lagrelii* represents the female; the sex dimorphism would be the same as in other large bears (see Kurtén, 1955). Several European "species" of *Indarctos* may represent sexual and individual variants.

Important literature includes Pilgrim (1931) and Zdansky (1924).

status. In the figures for the latest Pleistocene, all the Recent species of table 6.4 were also included, though in many cases their presence in these areas has not been established.

Simpson (1947) has shown that the formula

$$\frac{100C}{N_1},$$

Table 6.3.
Carnivores (Canidae, Ursidae, Hyaenidae, and Felidae) of the European and Chinese Cromerian

	Europe	China
Canidae	*Cuon priscus*	*Cuon alpinus antiquus*
	Canis lupus mosbachensis	*Canis lupus variabilis*
		Canis cyonoides
	Vulpes vulpes angustidens	*Vulpes vulpes* subsp.
		Vulpes corsac
	Alopex sp.	
	Lycaon lycaonoides	
		Nyctereutes megamastoides
Ursidae	*Ursus arctos deningeri*	*Ursus arctos* subsp.
	Ursus thibetanus schertzi	*Ursus thibetanus kokeni*
Hyaenidae	*Hyaena perrieri mosbachensis*	
	Hyaena brevirostris brevirostris	*Hyaena brevirostris sinensis*
	Crocuta crocuta subsp.	*Crocuta crocuta ultima*
Felidae	*Homotherium crenatidens crenatidens*	*Homotherium crenatidens ultimum*
		Megantereon inexpectatus
	Felis leo	
		Felis tigris
	Felis pardus	*Felis pardus*
	Felis lynx issiodorensis	*Felis ?lynx teilhardi*
	Felis silvestris	
		"Felis "microtus"
		Acinonyx sp.

Notes: On the Hyaenidae, see Kurtén (1956). Full data on the Ursidae will be published elsewhere; see also Erdbrink (1953). The Canidae have been revised recently by Thenius (1954). The Cromerian *Homotherium* can hardly be distinguished specifically from the Villafranchian *H. crenatidens,* of which it is a direct descendant; and this, in turn, cannot be validly separated from *H. ultimum. Felis teilhardi* is tentatively assigned to *F. lynx.*

Important literature includes Kretzoi (1937), Pei (1934), Zdansky (1925, 1928), and Zeuner (1937).

in which C is the number of species common to both faunas, and N_1 the number of species known in the smaller of the two faunas (thus, the percentage of common species in the smaller fauna), is the best simple measure of taxonomic resemblance. (Because of the denominations in table 6.5, we have to substitute N_2 for N_1 in one instance, for the early Pliocene, where $N_1 > N_2$).

These percentages are of the same order of magnitude throughout, indicating that patterns of distribution have been stable. The Recent percentage of common forms is somewhat lower; this may be the result of the activity of man.

Table 6.4.
Carnivores (Canidae, Ursidae, Hyaenidae, and Felidae) in the Recent Faunas of Europe and China, with Inclusion of Latest Pleistocene Forms (P)

	Europe	China
Canidae	*Cuon alpinus* (P)	*Cuon alpinus*
	Canis lupus	*Canis lupus*
	Canis aureus	
	Alopex lagopus	*Vulpes vulpes*
	Vulpes vulpes	*Vulpes corsac*
	Vulpes corsac	*Nyctereutes procyonides*
	[*Nyctereutes procyonides*][1]	
Ursidae	*Ursus arctos*	*Ursus arctos*
		Ursus thibetannus
	Ursus spelaeus (P)	
Hyaenidae	*Crocuta crocuta spelaea* (P)	*Crocuta crocuta ultima* (P)
	Hyaena hyaena monspessulana (P)	*Hyaena hyaena hyaena*
Felidae	*Felis silvestris*	
		Felis bieti
		Felis chaus
		Felis manul
	Felis lynx	*Felis lynx*
		Felis temmincki
		Felis bengalensis
		Felis nebulosa
	Felis pardus "spelaea" (P)	*Felis pardus* subsp.
	Felis leo spelaea (P)	
		Felis tigris

[1] Introduced.

Notes: Data for living species are from Ellerman & Morrison-Scott (1951). On the Hyaenidae, see Kurtén (1956). The name *Felis pardus spelaea* is invalid by homonymy with *F. leo spelaea*.

Table 6.5.
Number of Species of Canidae, Ursidae, Hyaenidae, and Felidae in Contemporary Deposits of Europe and China (N_1, N_2), and Number of Species Common to Both (C), from Tables 2–4; with Comparisons of Relative Frequencies of Common Species

	Europe (N_1)	China (N_2)	Common (C)	$\dfrac{100\,C}{N_1}$ or $\dfrac{100\,C}{N_2}$	$\dfrac{100\,C}{N_1 + N_2 - C}$
Recent	8	17	4	50	19
Late Pleistocene	14	18	9	64	39
Cromerian	15	17	9	60	39
Early Pliocene	17	12	7	58	32

Another formula,

$$\frac{100C}{N_1 + N_2 - C}$$

may also be used for such a comparison, but its use presupposes that we have complete knowledge of the two faunas to be compared, which is improbable for all except the Recent faunas. The fossil records, however, are fairly good, especially for the Pleistocene. There is a sharp drop from the Pleistocene to the Recent in this measure, resulting from the extinction of *Cuon alpinus* and the hyenas in Europe. The percentages for the fossil faunas are again approximately constant, and the consistent trend in the two measures indicates that the results are valid.

It may thus be concluded that the species ranges of larger carnivores have been comparable to those in the present day as far back as in the Pliocene and thereafter; if anything, they may have averaged slightly greater than today.

Both China and Europe belong to the same great faunal region. For the sake of completeness some comparison should also be made between regions nowaday characterized by markedly different faunal facies. I have selected the Indian and European faunas of the Pontian, the Villafranchian, and the present day.

The Siwaliks equivalent of the Pannonian, or Pontian sensu lato (Vallesian + Pikermian of the Spanish succession) is the Chinji + Nagri (Colbert 1935). These contain at least 12 species of the Canidae, Hyaenidae, Ursidae,[1] and Felidae. Of these, only one *(Crocuta eximia latro)* can at present be identified with any European species.[2] The possible identity of *Amphicyon polaeindicus* (with the synonyms *A. sindiensis, A. pithecophilus*) with a contemporary European ally cannot be excluded, but this must await a revision. The frequency of common species would thus be 8 percent.

Of 13 species present in the Pinjor zone of the Siwaliks, one *(Hyaena brevirostris neglecta)* appears to be conspecific with a European Villafranchian form, the nominate subspecies (Kurtén 1956). Another, *"Sivafelis" brachygnathus,* seems likely to be conspecific with the European Villafranchian cheetah, *Acinonyx pardinensis,* according to Matthew (see Colbert 1935). The frequency of common species may thus be around 15 percent.

In the present-day fauna of India, 23 species of these carnivorous families are represented. Four of them *(Canis lupus, Canis aureus, Felis lynx, Ursus arctos)* are identical with European species. The percentage 100 C/N_1 is thus 50, as in the comparison between Europe and China. A continued rise in the frequency of common species from the Pliocene to the Recent is indicated.

Application of the other measure, 100 $C/N_1 + N_2 - C$, gives roughly the following percentages: Pontian 4; Villafranchian 10; Recent 15. The trend is similar.

This might mean that species ranges have extended gradually during the period covered; or it may simply mean that the isolation of the Indian region has been gradually reduced—thus a special case without real bearing on the problem under consideration. In view of the apparent constancy of the European/Chinese indices for the same period, the latter interpretation appears more probable.

Similar comparisons of earlier Tertiary faunas would be of great interest, but they are beyond my competence. The tendency to create new species simply because of great distances between different localities will too often confuse the issue, and adequate revision of much of the material is necessary before valid comparisons can be made. From the fact that climatic zonation seems to have been less pronounced in the early Tertiary than in the Neocene, Pleistocene, and postglacial, it may perhaps be predicted that species ranges will be found to have averaged as great in the early Tertiary as now, or greater, and that distant localities will have a high frequency of common species.

In such revision the guiding principle should be the biological species concept, by means of population analysis, and the null hypothesis of the statistician. Preoccupation with analysis, at the cost of synthesis, has been prevalent since the turn of the century, and has sometimes led to almost fantastic feats of misinterpretation. Schindewolf's (1954) diatribe against "die alles vernünftige Mass übersteigende Arten-, Gattungs- und Namen- macherei, die vielfach als reiner Selbstzweck erscheint," is timely, and his espousal of the "new systematics" with its synthetic approach is laudable. A desirable result would be the reinstating of the species as the basic taxonomic category in paleontology, instead of the genus, which now masquerades in that role—often enough because it is actually doing the work of the species or the subgenus.

PLEISTOCENE PROBLEMS

Pleistocene stratigraphy, insofar as it is based on mammalian faunas, to a great extent relies on the migration phenomenon. The species is too long-lived to be used as a horizon marker when the stratigraphic zones represent much shorter time spans than the average longevity of species. Relative level of evolution within the scope of a species may be used over short geographic distances, but long-range correlation on this basis may be spurious. Geographic subspeciation may completely simulate temporal subspeciation, and thus a form may have a more advanced facies in one area than in another, in spite of perfect contemporaneity. As an example may be noted that the subfossil badgers *(Meles meles)* of Denmark are similar to the living Swedish badgers, whereas the living Danish population is of a more progressive type, with much lengthened crushing talonid in M_1 (Degerbøl 1933; Kurtén, 1954a).

Instead, correlation within type areas is based on a combination of this criterion with that of migration. The main interest is directed to the type of migration which is dominated by secular shifts in climate, and the faunas are interpreted as representing cold or warm stages. Correlation between distant areas with the same type of climatic history, say Europe and North America, is based on the assumption that the climatic oscillations were synchronous in both. But between glaciated and periglacial areas on one hand, and areas with a pluvial-dry oscillation on the other, the correlation becomes increasingly uncertain.

The only faunal event which may offer a solution seems to be the unchecked spread of a new, highly ubiquitous species into both areas. From the previous discussion it is clear that we may most hopefully look to the larger carnivorous forms to supply such evidence. One such instance, which furnishes a middle Pleistocene datum line, will be discussed here.

The Brevirostris-Crocuta Transition

In the Cromerian, or Gunz-Mindel interglacial, the spotted hyena *(Crocuta crocuta)* replaces its ecological predecessor, *Hyaena brevirostris,* in Europe. The same transition occurs in Asia, and moreover records from

Africa show *C. crocuta* to have been a Pleistocene immigrant on this continent. Details of the history are given in Kurtén (1956), and only a brief review, with some additional information, is necessary here.

Hyaena brevirostris Aymard is a gigantic member of the genus *Hyaena,* fairly closely related to the living *H. brunnea.* It occurs in the Villafranchian and Cromerian of Europe; in the Pinjor zone of the Siwaliks; in the Villafranchian and middle Pleistocene of China; and in the Djetis zone of Java.

Crocuta crocuta (Erxleben), the living spotted hyena, had a very great range of distribution in the Pleistocene, seemingly not much inferior to the territories of the living *Ursus arctos, Canis lupus,* and *Vulpes vulpes.* In Europe the species makes its first appearance in the Cromerian; in China it appears first in the middle Pleistocene of Choukoutien and correlative localities. The earliest record in Africa is from the australopithecine site of Swartkrans.

The probable ancestor of the spotted hyena is *Crocuta sivalensis* (Falconer and Cautley) from the Pinjor zone of the Siwaliks, India.

Both in Europe and in China the immigration of *Crocuta crocuta* is associated with the extinction of *Hyaena brevirostris.* Some conclusions about the nature of this replacement seem to be warranted.

The replacement of one species by another, ecologically related, species may occur in several ways:

(1) The extinction of the earlier form has no causal connection with the immigration of the later form. Both events result from the action of other factors, for instance climatic. This type of replacement is common in the Pleistocene.

(2) The extinction of one species permits the subsequent immigration of another.

(3) The immigrating form is adaptively superior to the local form, and ousts it through competition.

These possibilities may be combined in various ways. They have very different stratigraphic value. It is clear that many types of replacement depend entirely on local factors and thus are unlikely to provide tools for long-distance correlation between localities with different facies. What we require is a replacement of type (3), with the qualification that the migrant should be superior to the local form in all the realized environments, as well as capable of rapid spreading. It is suggested that the *brevirostris-crocuta* replacement is of this type.

The actual history of the replacement in a local fauna is best recorded at Choukoutien (Pei 1934). *Hyaena brevirostris* is confined to the deposits below the Peking Man level; *Crocuta crocuta* to those above that level. It is possible that the two forms existed together for a short time in this area; it is improbable that they did so for a prolonged period.

The correlative Chinese deposits show that both hyenas were at this time associated with an otherwise homogeneous fauna (the *Ailuropoda-Stegodon* fauna). This fauna covers a great range from Indochina to north-eastern China, and there is, of course, some regional differentiation, but that is all. There is no regional replacement of the two hyenas; they occur at random, one species or the other, scattered throughout this wide area. The replacement does not thus appear to be associated with any environmental change, and the suggestion is clearly that it was an internal affair between the two hyaenids; that *Crocuta* replaced *Hyaena* by competition.

In Europe the same replacement is enacted in a totally different setting. Apart from a few wide-ranging species, the fauna is quite distinct, and the environment probably highly different, from that in east Asia. The replacement of *Hyaena brevirostris* by *Crocuta crocuta* is recorded from the Forest Bed in East Anglia (Zeuner 1937), Süssenborn in Germany (Soergel 1936), and Gombaszög in Hungary (Kretzoi 1937). These localities are comparatively close to each other and must be considered as belonging to a single narrowly defined faunal province; there was at that time ample land connection between them. The difference in time between the immigration of *Crocuta crocuta* in Hungary and in East Anglia cannot reasonably have amounted to more than a few centuries (more probably a few decades) which, geologically speaking, is a most gratifying precision.

This evidence necessarily overrules indications to the contrary from relative level of evolution in small mammals, which may show geographic subspeciation within the range, and small differences in faunal composition reflecting local biotopes. Supposed differentiation among larger mammals appears to be spurious. The brown bears from these three localities, for instance, have been placed in different species: Soergel (1926) erected the "species' *Ursus sussenbornensis* for the Sussenborn form;

Andrews (1922) based another "species," *Ursus savini,* on the Forest Bed sample; while Kretzoi (1937) placed the Gombaszog bear in *Ursus eltruscus* as a distinct subspecies. The differences are imaginary, and all these populations belong in the single subspecies *Ursus arctos deningeri* von Reichenau, which is also known from other localities of Cromerian age.

The Forest Bed underlies the dichotomous Lowestoft (= Mindel) glaciation. Since the *brevirostris-crocuta* replacement apparently occurs within the Forest Bed, it must antedate Mindel I. The Süssenborn horizon with *H. brevirostris* and *C. crocuta* is regarded as representing the oncoming Mindel I or II; the former alternative must clearly be accepted. This fixes the time of the replacement within rather narrow boundaries; it must have occurred toward the end of the Cromerian interglacial, shortly before the beginning of Mindel I. Zeuner (1937) records *C. crocuta* from the Cromer Forest Bed, which is thought to be early Cromerian. I have not seen any specimen from there; those handled by me are *C. crocuta* from Palling and West Runton, and *H. brevirostris* from Palling and Mundesley; moreover a fine jaw of *H. brevirostris* from Bacton in a private collection, of which Dr. E. Ellis (Norwich) has kindly sent me a photograph. Some excellent teeth of *C. crocuta spelaea* from Corton Cliff may originate from the Corton Sands, which postdate the Cromerian (this was suggested by Dr. R. West, personal communication).

The faunas from Mauer and Mosbach, which would seem to represent a central level in the Cromerian, lack *C. crocuta;* the hyena is a late survival of *H. perrieri.* If the record from Cromer is correct, it may indicate that the sequence there may actually span the whole interglacial, and not only the earlier part of it.

Since *Hyaena brevirostris* is known from Europe, China, India, and Java, it is clear that the species was adaptable to many different climates and biotopes. It seems therefore that biotic factors would be more important in its extinction than abiotic. Furthermore, as far as we know, there was no common abiotic factor in all these areas; but a biotic change common to them all, except Java, is the imagination of the prospective competitor *Crocuta crocuta.* The only reasonable interpretation is that the giant *Hyaena* was ousted by the immigrant.

We do not know what advantage the spotted hyena may have had over *H. brevirostris.* But the spread of *C. crocuta* occurred subsequent to an evolutionary change from a more primitive to a more specialized condition. The probable ancestor of the species, the Villafranchian *C. sivalensis,* had a larger M^1 and a longer talonid in M_1. The reduction of these elements eliminates the chopping part of the bite and extends the slicing part. This improvement may seem ridiculously small to account for so decisive a change in the relation between two species; but it may have been sufficient to tip the scales and change the pattern of interspecific selection. Even very small selective advantages have been shown by population geneticists to be of far-reaching importance. It would appear that *Crocuta sivalensis* was able to hold its own in competition with *Hyaena brevirostris* in the Indian region, in Villafranchian time, but not outside of that local biotope; whereas its more progressive descendant attained superiority and migrated out into the Palearcitc and Ethiopian regions, ousting and supplanting *Hyaena brevirostris* throughout the area of that species.

ACKNOWLEDGEMENTS

My thanks are due to Dr. Joakim Danner, Helsingfors, Dr. E. A. Ellis, Norwich, and Dr. Richard West, Cambridge, for information pertinent to the present theme.

NOTES

1. The Ursidae are not represented in the Pontian, but they are present in later faunas. On Siwalik faunas, see Pilgrim (1932), Colbert (1935).

2. *Crocuta mordax* Pilgrim is a synonym of *C. "gigantea" latro* Pilgrim, as has also been suggested by Colbert (1935). The relationships of the middle Siwaliks hyenas are too complex to be discussed here; a revision is in progress.

REFERENCES

Andrews, C. W. 1922. Note on a bear (*Ursus savini sp.* n.) from the Cromer Forest-Bed: *Ann. Mag. Nat. Hist.*, London, 9:204–207.

Bader, R. S., 1955. Variability and evolutionary rate in the oreodonts: *Evolution* 9:119–140.

Colbert, E. H. 1935. Siwalik mammals in the American Museum of Natural History: *Am. Philos. Soc., Trans.*, new ser., vol. 26.

Degerbøl, M. 1933. Danmarks pattedyr i fortiden i sammenligning med recent former (with an English summary); Vid. Meddel. *Dansk Nat. hist. Forening*, 96(2):357–641, pl. 12–24.

Ellerman, J. R. and T. C. S. Morrison-Scott. 1951. Checklist of Palaeractic and Indian mammals, 1758 to 1946: British Mus. (Nat. Hist.), p. 1–810.

Erdbrink, D. P. 1953. *A Review of Fossil and Recent bears of the Old World.* Deventer.

Kalela, O. 1940. Uber die Einwanderung and Verbreitung des Iltis, *Putorius putorius* (L.), in Finnland: *Ann. Acad. Sci. Fennicae* (ser. A) 54:(6):1–76.

Kalela, O. 1948. The occurrence of roe deer in Finland and changes in its distribution in the adjoining areas (in Finnish, with an English summary): *Suomen Riista* (1948/3):34–56.

Kalela, O. 1949. Uber Fjeldlemming-Invasionen und andere irreguläre Tierwanderungen: *Ann. Zool. Soc. "Vanamo"* 13,(5):i–iv, 1–90.

Kretzoi, M. 1937. Die Raubtiere von Gombaszög nebst einer Ubersicht der Gesammtfauna: *Ann. Mus. Nat. Hungarici, pars min. geol. paleont.*, 31:88–157, pl. 1–3.

Kurtén, B. 1954a. Observations on allometry in mammalian dentitions; its interpretation and evolutionary significance: *Acta Zool. Fennica* (85):1–13.

Kurtén, B. 1954b. The type collection of *Ictitherium robustum* (Gervais ex Nordmann) and the radiation of the ictitheres: *Acta Zool. Fennica* (86):1–26.

Kurtén, B. 1955. Sex dimorphism and size trends in the cave bear, *Ursus spelaeus* Rosenmuller and Heinroth. *Acta Zool. Fennica* (90):1–48.

Kurtén, B. 1956. The status and affinities of *Hyaena sinensis* Owen and *Hyaena ultima* Matsumoto: *Am. Mus. Novitates* (1764):1–48.

Orlov, J. A. 1941. Tertiary Carnivora of West Siberia. IV. Hyaeninae; *Trav. Inst. Paleont. URSS* 8, (3):40–60, 5 pl.

Pei, W. C. 1934. On the Carnivora from locality 1 of Choukoutien: *Paleont. Sinica, ser. C,* 8 (1):1–116, pl. 1–24.

Pei, W. C. 1939. Geochronological table no. 1, an attempted correlation of Quaternary geology, palaeontology and prehistory, in Europe and China: Univ. London, Inst. Arch Occ. Pap., no. 2, p. 1–16.

Pilgrim, G. E. 1931. Catalogue of the Pontian Carnivora of Europe in the department of geology: p. i–vi, 1–174, 2 pl., Brit. Mus. (Nat. Hist.).

Pilgrim, G. E. 1932. The fossil Carnivora of India: *Palaeont. Indica,* new ser., 18:1–232, pl. 1–10.

Schindewolf, O. H. 1954. Zur taxonomie rezenter und fossiler Organismen: *C. R. Congr. Geol. Int. Alger* (1952) (19):81–91.

Simpson, G. G. 1947. Holarctic mammalian faunas and continental relationships during the Cenzoic: *Geol. Soc. Am., Bull.* 58:613–688.

Simpson, G. G. 1953. *The Major Features of Evolution.* New York: Columbia University Press.

Soergel, W. 1926. Der Bär von Süssenborn. Ein Beitrag zur näheren Kenntnis der diluvialen Bären: *Neues Jahrb. ser. B,* 54:115–156, pl. 3–14.

Soergel, W. 1936. *Hyaena brevirostris* Aymard und *Hyaena* ex aff. *crocotta* Erxl. aus den Kiesen von Süssenborn: *Zeitschr. Deutschen Geol. Gesell.* 88:525–539, pl. 45–48.

Thenius, E. 1951. Zur odontologischen Charakteristik von *"Felis" leiodon* aus dem Pont von Pikermi (Griechenland): *Neues Jahrbuch* (1951/3): 88–96.

Thenius, E. 1954. Die Caniden (Mammalia) aus dem Altquartär von Hundsheim (Niederösterreich) nebst Bemerkungen zur Stammesgeschichte der Gattung *Cuon: Neues Jahrbuch,* 99(2):230–286.

Zdansky, O. 1924. Jungtertiäre Carnivoren Chinas: *Palaeont. Sinica,* ser. C., 2 (1):1–155, pl. 1–33.

Zdansky, O. 1925. Quartäre Carnivoren aus Nord-China: *Palaeont. Sinica,* ser. C., 2 (2):1–30, pl. 1–4.

Zdansky, O. 1928. Die Säugetiere der Quartärfauna von Chou-k'ou-tien: *Paleont. Sinica, ser. C,* 5 (4):1–146, pl. 1–16.

Zeuner, F. E. 1937. A comparison of the Pleistocene of East Anglia with that of Germany: *Prehist. Soc. Cambridge, Proc.,* new ser., 3:136–157.

SEVEN

Holarctic Land Connections in the Early Tertiary

SINCE THE PUBLICATION of Simpson's (1947) careful study of the Holarctic mammalian faunas and continental relationships during the Cenozoic, it has been generally accepted that the Bering bridge was the main or only route of intermigration between the Nearctic and Palaearctic regions. "Evidence is inadequate for the earliest Tertiary, but does not favor any other route, and from the late Eocene on increasingly strong evidence points to the Bering route" (p. 686). It is the purpose of the present paper to show that the evidence for the early Tertiary, which is now more adequate than when the above was written, supports the older idea of a direct, probably North Atlantic, route in the Paleocene and early Eocene.

The geological evidence indicates that Europe and Asia were separated from each other for most of the early Tertiary by broad tracts of water (fig. 7.1). If Europe communicated with North America by way of the Bering bridge only, it must be assumed that the mammals also crossed the Turgai strait.

Migration is a somewhat misleading word for the spread of mammals from one continent to another. It suggests coordinated movement toward a goal as in the various instances of seasonal migration. But here it should only be understood to mean ordinary population spread. The land bridges did not function as highways; they were simply areas that became populated by various mammal species, as they emerged from the water and became covered with vegetation.

Reprinted, with permission, from *Commentationes Biologicae* (1966) 29 (5):1–5.

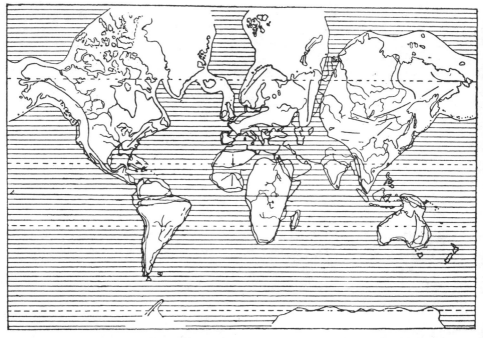

Fig. 7.1. Palaeogeography of the Eocene, after Schaffer. The system of Holarctic land bridges indicated is consistent with the faunal history in the Paleocene and early Eocene.

In this sense, Asia might be regarded as a land bridge between Europe and North America, provided that the faunal exchange was routed along the Bering bridge. The faunal composition of a functioning land bridge may be predicted on the basis of the bridgehead faunas. The common element should be strongly dominant in the fauna. There should also be a number of European forms that spread east onto the bridge but not across it, and similarly a number of American forms that did not spread all the way to Europe. Finally, there should be an endemic component in the fauna, but under the given circumstances this element should be weak and mostly concentrated in marginal areas.

The actual fossil faunas of Asia in the late Paleocene and early Eocene are utterly different in nature from what would be predicted by the land bridge theory. They have a large endemic component, and are also very strongly influenced by North America, while the European element is insignificant. The most important fact is that the American-European migrant element, which should be dominant, is lacking almost entirely.

Looking first at the Gashato assemblage, which represents the late Paleocene fauna of Asia in a region close to the Bering bridge, we may note an important endemic element. The lagomorph, the insectivores, the condylarth and probably all the creodonts are probably not only endemic but also autochtonous, with a long history in Asia behind themselves. There is also a large number of forms with a North American relationship: the taeniolabiid multituberculates, the pantodonts, the dinocerates, and notoungulates. There is nothing definitely European in this fauna.

This means that all of the forms known to be common to Europe and North America are lacking. They are such as the plesiadapid and carpolestid primates; the ptilodont multituberculates; the hyopsodontid and meniscotheriid condylarths; the genus *Arctocyon* (= *Claenodon*). These forms represent the major part of the European late Paleocene fauna. They must have formed an important element in the land bridge fauna too. Their absence at Gashato might be explained in terms of some natural obstacle that rerouted the Bering migrants, for instance through southern Asia; but the explanation is negated by the fact that Gashato does contain an important Bering bridge fauna and thus could not have been isolated from the migration route.

The situation in the Paleocene cannot be reconciled with the idea of a Bering bridge connection between Europe and North America, and the evidence for the lower Eocene is even stronger. At this stage the similarity between the European and North American faunas is so great that, in the words of Simpson (1947), the two continents "were zoogeographically essentially a single region." Numerous genera were identical in both areas, for instance *Phenacodus, Coryphodon, Hyracotherium, Esthonyx, Ectoganus, Paramys, Dissacus, Pachyaena, Palaeonictis*; and some of the species are probably also identical. The interchange concerned the major part of the faunas and must have required a broad corridor with a varied environment suitable for many different species. The land bridge must have been completely dominated by this common Europeo-American fauna.

The early Eocene of Asia is not too well known but the Ulan Bulak and some other faunules give some idea of its mammals (Thenius 1959). They show a very strong American influence but the American elements are almost consistently different from those that appear in Europe. There is a member of the Barylambdidae, otherwise only found in the Paleocene of North America (Simons 1960). *Mongolotherium* is a dinocerate with American affinities, and the same affinities are indicated by the tapiroids *Heptodon* and *Homogalax* recently found in China (Chow and Li 1965); both genera also occur in North America. In contrast, the typically European lophiodont tapiroids are missing; Asiatic forms previously regarded as belonging to this family are now regarded as members of an endemic family (Lophialetidae; Radinsky 1965). The creodont *Mesonyx* should also be noted; it was later to appear in North America but did not get to Europe. The pantolestid *Palaeosinopa* is the only form common to Europe, Asia, and North America, but it can hardly be called a dominant element. It may have crossed the Turgai strait, but it is perhaps more probable that it used the same direct route as the other interchange between Europe and North America, as well as the Bering bridge.

The American influence on the Asiatic fauna continued in the late Eocene, but at this time some faunal interchange between Asia and Europe also occurred. A temporary regression is assumed to have created a land bridge across the Turgai strait. There is also some interchange between Europe and North America. This time it is quite striking that the common European and North American elements also tend to appear in Asia. This is true for *Pterodon, Hyaenodon,* and *Miacis.* The middle to late Eocene migration between Europe and North America was quite moderate and it seems probable that only the Bering route was involved, although island-hopping along the remnants of the direct route may also be involved.

The following conclusions may be suggested:
There was no unified Palaearctic zoogeographic region in the Paleocene and early Eocene. The Holarctic region of the present day was then divided into three fully separate regions, the European, the Asiatic, and the Nearctic. The history of their interconnections may be summarized as follows:

(1) The connection between Europe and Asia was broken from the (late) Paleocene up to some time in the middle or late Eocene when some intermigration took place.

(2) Intermigration between Asia and North America was fairly strong, with probably at least three separate phases, one in the late Paleocene, another in the early Eocene and a third at some time in the late middle Eocene or beginning of the late Eocene.

(3) Intermigration between Europe and North America, by a direct route not crossing Asia, was marked in the late Paleocene and very intense in the earliest Eocene, but then ceased. At most it might have been resumed at a moderate scale between the middle and late Eocene (island-hopping?) but the evidence is insufficient.

The positions of two of the land bridges seem to be clearly indicated. (1) Bridge across the Turgai strait between Europe and Asia; (2) Bering bridge between Asia and North America. The position of (3) is less certain but the most probable one is evidently a North Atlantic connection including the British Isles, Iceland, and Greenland, which is found in most paleogeographic maps (fig. 7.1).

REFERENCES

Chow, Minchen, and Li, Chuan-Kuel 1965. *Homogalax* and *Heptodon* of Shantung. *Vertebrata Palasiatica* 9:19–22.

Radinsky, Leonard B. 1965. Early Tertiary Tapiroidea of Asia. *Bull. American Mus. Nat. Hist.* 129:181–262.

Simons, Elwyn L. 1960. The Paleocene Pantodonta. *Trans. American Philos. Soc.*, new ser., 50, (part 6):1–100.

Simpson, George Gaylord 1947. Holarctic mammalian faunas and continental relationships during the Cenozoic. *Bull. Geol. Soc. America* 58:613–688.

Thenius, Erich 1959. *Tertiär, II*. Teil: Wirbeltierfaunen. Enke, Stuttgart.

EIGHT

Continental Drift and the Palaeogeography of Reptiles and Mammals

INTRODUCTION

THE STUDY OF continental drift has recently been revitalized by the research on palaeomagnetism, and there can now be little doubt that drift has in fact occurred (Runcorn 1961; Irving 1964). In the early Mesozoic, the continents of the earth were still gathered in two supercontinents, Laurasia in the north and Gondwanaland in the south, separated by the Tethys sea. Of the two, Gondwanaland was the largest, incorporating the land masses that now form South America, Africa, Antarctica, the Indian peninsula, and Australia; while Laurasia consisted of North America, Europe, and Asia north of the Tethys. Redistribution of the continents commenced in the Mesozoic, and a phase of comparatively rapid drift was enacted in the Cretaceous and early Tertiary. Laurasia split up into North America and Eurasia north of the Tethys. The five southern continents drifted apart, the Indian peninsula being finally added to the land mass of Asia.

There has been little serious study of the effects of these concepts on animal palaeogeography and the evolution of the Tetrapoda. Yet it is obvious that palaeogeography is of basic importance in this respect. There is, for instance, the possibility that land vertebrates arose from

Reprinted, with permission, from *Commentationes Biologicae* (1967) 31 (1):1–8.

quite different stocks of crossopterygians in Laurasia and Gondwanaland. A diphyletic (or polyphyletic) origin of tetrapods has in fact been long advocated by Jarvik (1942).

The present contribution is a comparison between the radiation of the reptiles (late Carboniferous to Cretaceous) and that of the mammals (late Cretaceous and Cenozoic). These two radiations took place in radically different geographic settings. The reptilian radiation occurred while the land masses of the earth were collected into two enormous blocks, the mammalian while they were split into fragments.

At least 33 mammalian orders have been in existence during the 100 million years or so of late Cretaceous and Cenozoic time, so that the relationship between ordinal variety and time is 1:3 for the mammals. The Age of Reptiles lasted more than 200 million years, and in that time about 20 orders were produced, so that the relationship is 1:10.

It may, of course, be maintained that reptilian and mammalian orders are not exactly equivalent. Perhaps it would be more correct to use the reptilian suborder instead, in which case the relationship becomes 1:5 approximately. The mammals still come out superior in the rate of radiation.

Can this be correlated with the greater fragmentation of the continents during the mammalian radiation? What is in fact the relationship between numbers of continents, sizes of continents, and production of tetrapod orders? To answer these questions, we must attempt to determine the geographic origin of the reptilian and mammalian orders.

THE REPTILIAN RADIATION

The probable origin of the reptile orders—in Gondwanaland or Laurasia—is very hard to ascertain in most cases. The fossil record of the earliest reptiles (Carboniferous and early Permian) is so scanty in Gondwanaland that it would be hazardous to draw any conclusions as regards the Cotylosauria, the Pelycosauria, and the Protorosauria. There seems also at present to be little evidence on the probable origin of the Ichthyosauria, Thecodontia, Crocodilia, and Saurischia. As regards the other orders, the following tentative suggestions may be brought forth.

A Gondwanaland origin may perhaps be indicated as regards the following orders:

Chelonia. This is dependent on the interpretation of the Gondwanaland *Eunotosaurus* as an early representative.

Mesosauria. Restricted to Gondwanaland.

Eosuchia. Southern distribution in the Permian, invade Laurasia much later.

Rhynchocephalia. First appearance in Gondwanaland, later in Laurasia.

Ornithischia. First ornithischians in the late Trias of Gondwanaland; the order invades Laurasia in the early Jurassic.

In addition, a Gondwanaland origin may possibly be suggested for the Ictidosauria, though the comparatively poor fossil record of this order makes any conclusions very uncertain. It might perhaps also be thought that the sudden appearance of full-fledged pterosaurs in the lower Jurassic of Laurasia might reflect an immigration from an unknown evolutionary center somewhere in the south, but such a reasoning is obviously very tenuous. Still, even if these cases are left, out, there are at least five reptilian orders for which a Gondwanaland origin is indicated.

A Laurasian origin is possible in the following cases:

Sauropterygia. Nothosaurs appear in the early Trias of Laurasia, enter Gondwanaland somewhat later.

Therapsida. Laurasian forms transitional between Pelycosauria and Therapsida may indicate that the origin of this order lay in the northern continent.

Finally, an argument might be made for a Laurasian origin of the Squamata, based on the Rhaetic gliding kuhneosaurids.

Thus, among those comparatively few reptilian orders which can be assigned to one or the other of the supercontinents, five or six seem to have arisen in Gondwanaland, and two or three in Laurasia.

In the Permian and early Mesozoic, Gondwanaland was not only the largest land mass of its time, but also the one with the most varied climate (fig. 8.1). Known Permian and Triassic palaeoclimates of Gondwanaland

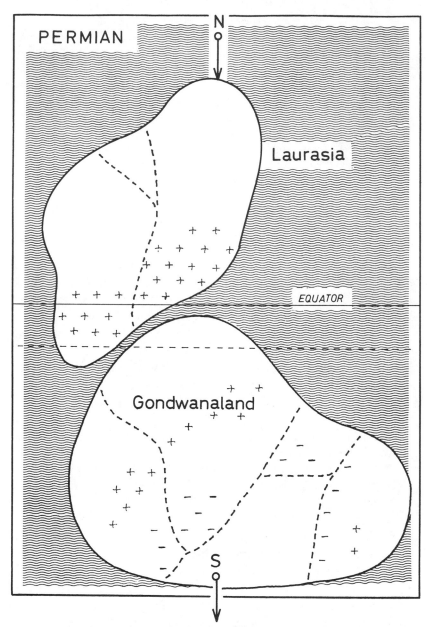

Fig. 8.1. Diagrammatic representation of Permian paleogeography as indicated by continental drift. Arrows indicate pole movements from early Permian to late Triassic; dashed line is late Triassic paleoequator. Plus signs indicate general distribution of warm-climate indicators (desert sands, evaporties, coral reefs, coal beds); minus signs, cold-climate indicators (tillites, glacial striae). Laurasia consists of North America, Greenland, and Eurasia (Europe in the south, Asia in the north). Gondwanaland is composed to South America, Africa, India, Australia, and Antarctica (with the South Pole; counting clockwise).

range from arctic (glacial) to tropical in nature, while the smaller Laurasian continent was mainly tropical and temperate. In fact, Gondwanaland might perhaps be compared with Eurasia of the present day, while the role of Laurasia was analogous to that of Africa or South America. Considering all this, it appears only natural that the greatest radiation of tetrapods should have occurred in Gondwanaland.

Intermigration between the two continents obviously was possible, but perhaps intermittent. The land bridge probably lay at the western end of Tethys.

THE MAMMALIAN RADIATION

The palaeogeographic situation at the time of the mammalian radiation (late Cretaceous and early Tertiary) was radically different. Both Gond-wanaland and Laurasia had split up into subcontinents, and in this phase the fragmentation was maximal. In addition, the great marine transgressions of the late Cretaceous and early Tertiary created large interior seas, dividing some of the remaining continental blocks into islands.

The separate continents were now as follows (fig. 8.2).

(1) North America had drifted away from Eurasia, but apparently retained contact, intermittently at least, by way of a land bridge across the North Atlantic (1–2). Animal geography indicates that this land bridge was in function up to and including the early Eocene, then foundered. In addition, another intermittent land bridge (the Bering bridge) extended from North America to Asia (1–3).

(2) Although the remaining part of Laurasia, Eurasia, remained a single continental block, it was long divided into a European and an Asiatic part by an interior seaway from Tethys to the Arctic, east of the Ural mountains. This appears to have caused a practically complete migration barrier between the land faunas of Europe and Asia, up to the middle Eocene when the sea began to retreat from this area.

(3) Asia, then, had little or no direct contact with Europe, except by way of a circular route crossing North America.

(4) South America, one of the daughter continents of Gondwan-aland, became almost completely isolated at the beginning of the

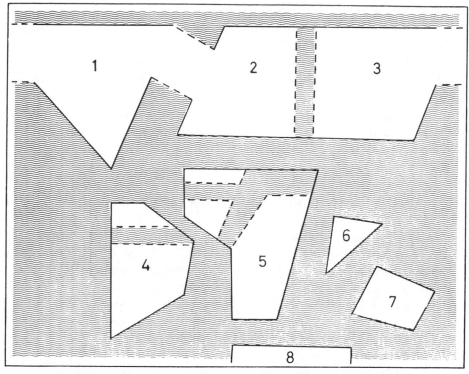

Fig. 8.2.   Palaeography at the Cretaceous-Tertiary transition. Land bridges and interior seas in dashed outlines. Areas are 1, North America; 2, Europe; 3, Asia; 4, South America; 5, Africa; 6, Indian peninsula; 7, Australia; 8, Antarctica.

Tertiary. On zoogeographic grounds, however, some Late Cretaceous intermigration with both North America and Australia must be postulated. It probably occurred in the form of "island-hopping" along a so-called sweepstakes route (Simpson 1953): the Gondwanaland remnants probably formed the stepping stones between South America and Australia. After this, South America remained isolated up to the later Tertiary.

(5) The African continent was an important evolutionary center. A large part of it was flooded by shelf seas, dividing the continent into islands. It was more or less completely isolated in the Paleogene, apart from some "filter-barrier" migration.

(6) The Indian peninsula forms another fragment of Gondwanaland, which drifted northward and finally came into contact with the

Asiatic land mass, piling up the Himalaya in the process. Its zoogeographic role in the early Tertiary is still somewhat obscure. There seems to have been some limited intermigration with Asia.

(7) Australia, though probably in indirect contact across Gondwanaland with South America in the late Cretaceous, has been completely isolated since then.

(8) The role of Antarctica in the evolution of the mammals is little known.

There were thus no less than eight separate nuclei, in various degrees of isolation from each other, to provide areas for mammalian radiation. Let us now consider how the evolution of the mammalian orders was distributed.

Europe, Asia and North America were not completely isolated from each other, and so it is difficult to determine the place of origin of the Holarctic mammalian orders. At present it is only in North America that we have a good fossil record for mammals from the late Cretaceus and the earlier Paleocene. Only in the case of those orders that appear a little later in the fossil record, inferences of this kind are now possible. Thus it might be suggested, for instance, that the Rodentia arose in North America, and the Lagomorpha in Asia. Otherwise, little can be said of the exact centre of origin for the long array of orders that probably arose in the Holarctic area—the Multituberculata, Insectivora, Deltatheridia, Dermoptera, Chiroptera, Primates, Carnivora, Condylarthra, Perissodactyla, Artiodactyla, Tillodontia, Taeniodontia, Amblypoda, Dinocerata, Pholidota, and possibly the Edentata, though it may also be of South American origin.

South America, an isolated island continent, produced a highly varied fauna with the orders Paucituberculata, Astrapotheria, Pyrotheria, Litopterna, Notoungulata, and perhaps Edentata: the South American origin of the two last-mentioned is less certain. The order Marsupicarnivora is distributed between South America and Australia, and of its origin nothing more can be said than that it apparently lay in some one of the daughter continents of Gondwanaland.

Africa, in semi-isolation, produced a large crop of orders; their preEocene history is unfortunately unknown. Here belong the Hyracoidea, Embrithopoda, Proboscidea, and probably also the Sirenia, Desmostyliformes, and Tubulidentata. Among infraordinal taxa of special interest with an African origin may be mentioned the pongid-hominid group, which appears to be represented by *Propliopithecus*.

In the Australian continent only two orders of mammals appear to have evolved, the Peramelina and Diprotodonta. But early Tertiary and Cretaceous Australian faunas are not yet known. Again, there are no orders that on present knowledge can be assigned to the Indian peninsula, or to Antarctica.

The geographic origin of Cetacea is unknown. The chances are that the whales descended from some Holarctic group of land mammals, but an African or Indian origin cannot at present be excluded.

Thus, the three Holarctic continents together produced some 17 or 18 orders, or on an average, about six orders each; Africa and South America also produced about six orders each; Australia two; India (and Antarctica) none, on present evidence. There is obviously a good correlation between size of area and number of orders produced, just as in the case of Gondwanaland and Laurasia.

From a zoogeographic point of view, the Tertiary was an almost continuous amalgamation process. At the beginning of the period, the eight continents were almost completely isolated from each other; then they gradually came into contact, land bridges were being formed, and the local faunas began to intermigrate. In this process there was a gradual weeding out of ecologically overlapping form, which in some instances led to the complete extinction of entire orders—this is especially marked in the South American fauna. Thus the great number of mammalian orders produced was also strongly influenced by the separation into many disjunct areas; otherwise this remarkable duplication of adaptive types could hardly have been brought about.

CONCLUSIONS

Study of the reptilian radiation in the Permian and Mesozoic, and the mammalian in the Cretaceous and Tertiary, in the palaeogeographic setting indicates by continental drift, suggests that the texonomic variety produced is correlated with the size of the area available, with its environmental variety, and with the number of disjunct areas.

REFERENCES

Dietz, R. S., & Sproll, W. P. 1966. Equal areas of Gondwana and Laurasia (ancient super continents). *Nature* 112:1196–1198.

Falcon, N. L. 1967. Equal areas of Gondwana and Laurasia. *Nature* 113:580.

Irving, E. 1964. *Paleomagnetism and Its Application to Geological and Geophysical Problems*. New York: Wiley.

Jarvik, E. 1942. On the structure of the snout of crossopterygians and lower gnathostomes in general. *Zool. Bidr. Uppsala* 21:235–675.

Runcorn, S. K., ed. 1961. *Continental Drift*. New York and London: Academic Press.

Simpson, G. G. 1953. *Evolution and Geography*. Condon Lectures. Eugene, Oregon: Oregon State Board of Higher Education.

ADDENDUM

The outcome of a recent discussion on the areas of Gondwanaland and Laurasia (Dietz and Sproll 1966; Falcon 1967) is that the supercontinents were semi-equal in size, Gondwanaland being slightly larger and with a more varied environment.

IV
DYNAMICS OF FAUNAL CHANGE

NINE

On the Longevity of Mammalian Species in the Tertiary

CONCEPT OF SPECIES HALF-LIFE

DIRECT MEASUREMENT of the length of life of a fossil species usually meets with many difficulties; an accurate absolute chronology and a very good fossil record are both necessary. As regards species of mammals, these two conditions are only fulfilled for part of the Quaternary period, and even here only approximately so. To help overcome the difficulties, I have attempted to devise another method of estimating the mean species longevity, by using the concept of half-life (Kurtén 1960). The theoretical basis is a model population of species with the following properties: (1) It is closed to migration; (2) it keeps constant size (constant number of species) through time; (3) the rise of new species and the extinction of old, or the faunal turnover, occur randomly distributed around a constant mean rate.

Actual species populations probably approximate these conditions to varying degrees. The importance of the deviations from the ideal may be briefly evaluated as follows.

(1) Almost no species population is closed to migration. As regards the subject of the present article, the Tertiary mammals, some continents have been almost perfectly isolated during some epochs, and for these epochs surveys of the total endemic faunas would fulfill the condition. No such survey is possible at present.

Reprinted, with permission, from *Commentationes Biologicae* (1959) 21 (4):1–14.

This deviation from the model, however, loses some of its importance when it is remembered that migration usually is correlated with evolution, insofar as the migrating species tends to adapt to the new environment and frequently does so by evolving into a new species; or else its migration may be due to its having attained an evolutionary stage that increases its adaptive potential. Thus, when a new species appears in a local fauna, it will probably quite often be "really" new, i.e., specifically distinct from its immediate ancestor. Disappearance from a local fauna is not proof of total extinction, but in some instances the population may have contrived to survive in other areas only by evolving into a new species. On the whole, the bias introduced by migration would seem to be somewhat slighter in the Tertiary, with its relatively stable environment, than in the Quaternary.

(2) Though the number of species in local faunas certainly does not remain constant, it is likely to be approximately so in the long run.

(3) There is reason to suppose that this condition is never exactly fulfilled. Deviations are of two types: (A) Nonrandom distribution of turnover rates around the mean. This deviation appears to be slight and unimportant, since fitted half-life curves appear to describe actual case-histories quite well (Kurtén, in the press, fig. 10). (B) Changes in mean turnover rates. Such changes are probably important (Kurtén, *loc.cit.*) and form one of the topics of this investigation.

In the model population, the history of all the species living at a point in time may be pictured as in fig. 9.1. Of the species that lived at time T, a few appear at a relatively early date; gradually, others are added, at the same relative rate (but at an absolutely accelerating rate). The curve depicting the rise of the temporal stratum of T will be a half-life curve; in a semilogarithmic diagram, with the species numbers (percentages) on the log scale, it will be a straight line. The half-life is the time during which the percentage rises from 50 to 100 (or from 25 to 50, from 12.5 to 25, etc.). Provided that the rate of turnover continues unchanged, the number of species living at time T will shrink in post-T times at a similar rate, the half-life again being the time of contraction from 100 to 50 percent.

The half-life is directly related to the mean longevity; the mean longevity is 2.89 times the half-life (see Kurtén, *op.cit.*).

Measuring species longevity by this method has some distinct advantages. In Tertiary mammalian faunas, species are normally recorded as present in one stage, or in two or more stages (or successive local faunas,

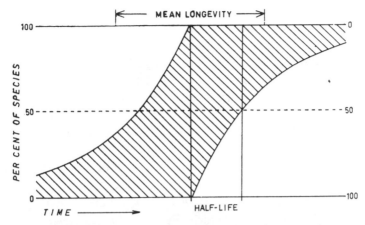

Fig. 9.1. Diagrammatic representation of the history of a temporal stratum (cross-hatched) in a model population of species.

etc.). There is generally no possibility at all to determine, even approximately, the longevity of species present in one stage only. The species recorded as present in two successive stages may actually be so throughout the two stages or only through part of them. Thus the percentual error of the raw data will still be very great. It is smaller for species living through three stages, but very few species are long-lived enough for that. Furthermore, records of part of the history of the species are certainly often missing.

The half-life method does not presuppose complete knowledge of the temporal range of the species. All that is needed is actually a succession of two (or more) fairly large and evenly recorded faunas. If, for instance, two successive age-stages, A and B, are considered, the half-life may be determined, in terms of the age unit, either from the percentage of B species in fauna A, or from that of A species in fauna B; or, preferably, as a weighted mean of both estimates. If three or more successive faunas are known, the results may be checked by counts of the percentages of C species in fauna A, A species in fauna C, and so on.

In such determinations a considerable variability may be expected, and it is actually found; it is probably real in part, and due to factor (3B) above (changes in mean turnover rates through time), but it is probably also, and perhaps mainly, due to incompleteness of the fossil record, chances of sampling, and incorrect taxonomy. To reduce the influence of sampling errors, the estimates in the following instances have been averaged, as weighted means, for relatively large spans of time, mostly representing a whole epoch or more.

There remain, then, three important factors of bias:

(4) Incompleteness of the record. This will tend to reduce the estimated half-life. The importance of this factor is unknown, and it is difficult to gauge, but it will play a part in any attempt at estimating mean species longevities; and the bias introduced by this factor is radically lower for half-life estimates than for direct computation of species longevity. A compensating factor may be that large populations, which are probably more conservative in evolution, will be better known than small ones. If two successive faunas are very unequal in known species numbers, the percentages in the smaller fauna will give a better estimate than those in the larger fauna (which will be too low).

(5) Incorrect taxonomy. The species concept of many paleontologists, especially as regards earlier work, is typological, and tends to result in over-splitting. This will reduce the apparent half-life. On the other hand, incompletely known forms are usually tentatively referred to a known species, which may tend to increase the apparent half-life.

(6) The absolute chronology is still rather vague, though the order of magnitude of the time corresponding to the Tertiary epochs is known. Different stages do not have the same absolute time equivalent, but by using, again, averages for a prolonged period, it appears that useful estimates may be obtained.

Keeping all these sources of bias in mind, it is clear that we can hope to obtain only very tentative values for the species longevity of Tertiary mammals. They will, however, give some idea of the order of magnitude; they will also permit tentative comparisons of evolutionary rates in different groups of mammals and during different periods. Finally, they will, I believe, give a definite answer to the question which prompted me to make this attempt: Were the evolutionary rates in the Tertiary different from those in the Quaternary? It seems to be clear that they were indeed different.

MATERIAL

The material used in this study is as follows:

(A) Spanish local faunas from the Burdigalian, Vindobonian, Vallesian, and Pikermian, as recorded by Crusafont (1958). These are all large and well-studied faunas. Included are the faunas from the Vallés-Penedés, the meseta, and the Calatayud-Teruel area. The Pannonian faunas of the last-mentioned areas seem in the main to be of Pikermian age, and have been so treated here.

(B) Siwaliks local faunas from the Chinji, Nagri, Dhok Pathan, and Tatrot and Pinjor (combined), as tabulated in Colbert (1935). These, too, are large and well-known faunas, but probably somewhat excessively elaborated on the species level (as noted by Colbert).

(C) Chinese local faunas from the Pontian, "middle" (= upper) Pliocene, "upper" Pliocene (= Villafranchian or lower Pleistocene), and "lower" (= middle) Pleistocene, as tabulated by Teilhard and Leroy (1942). In this instance Quaternary faunas were used, but the Villafranchian is a stage of Tertiary rather than of Quaternary type. The faunas are large and well-known.

In all these instances, emendations of the taxonomic treatment have been made on some points.

(D) American Paleocene faunas from Fort Union, Montana (Simpson 1937), and Tiffany, Colorado (Simpson 1935). The Fort Union middle Paleocene fauna is divided into three zones (1, 2, and 3 of Simpson 1937, table 3) of which the uppermost may be transitional to the upper Paleocene. In this case, therefore, the time unit is less than one age, and may be set approximately as one-half an age. The taxonomy is taken from Simpson without changes.

The Paleocene record is more spotty than the Neogene Old World records, but is of great interest, because it dates from an epoch in which the Mammalia are considered to have undergone "explosive" evolution: I have included it for this reason, though it represents a field in which I can claim no personal competence.

METHOD OF ANALYSIS

The method used in the analysis will be most readily understood if illustrated by an example, and I have selected the Spanish Mio-Pliocene Carnivora. Table 9.1A shows the basic distribution by first and last appearances; table 9.1B the cumulative distribution showing the total number of species belonging to different temporal strata, present at the given time. Thus, for instance, the Vindobonian fauna contains 22 species; 2 of them were already present in the Burdigalian; 4 of them

Table 9.1.
A, distribution, by first and last appearances, of species of Carnivora in
the Mio-Pliocene of Spain. B, cumulative distribution from A. Data
from Crusafont. Abbreviations of age-stage names: B, Burdigalian; Vi,
Vindobonian; Va, Vallesian; P, Pikermian.

		A					B			
		First appearances								
		B	Vi	Va	P		B	Vi	Va	P
Last	P	—	1	6	7	P	—	1	7	14
appearances	Va	1	2	8		Va	1	4	18	
	Vi	1	17			Vi	2	22		
	B	12				B	14			

survived to the Vallesian; and one survived to the Pikermian. The
Vindobonian fauna may thus be said to contain 9 percent species dating
from the Burdigalian or earlier: 18 percent that will survive to the
Vallesian or later; and 5 percent species that will survive to the Pikermian or later.

The average percentage of previous-stage species in a given fauna is
obtained as

$$\frac{100(2 + 4 + 7)}{22 + 18 + 14} = 24\%$$

and the average percentage of next-stage species as

$$\frac{100(2 + 4 + 7)}{14 + 22 + 18} = 24\%$$

The results happen to be identical in this case, but this is not always the
case. A weighted mean percentage is obtained from

$$\frac{100 \times 2(2 + 4 + 7)}{14 + 2(22 + 18) + 14} = 24\%$$

The half-life, expressed with the local age as a unit, is then

$$\frac{\log 100 - \log 50}{\log 100 - \log 24} = \frac{0.301}{0.620} = 0.49, \text{ or about } \tfrac{1}{2} \text{ age.}$$

Weighted mean percentages for temporal strata in faunas two stages apart are obtained in an analogous way:

$$\frac{100 \times 2(1 + 1)}{14 + 18 + 22 + 14} = 5.9\%$$

The half-life calculated on this basis is

$$\frac{2(\log 100 - \log 50)}{\log 100 - \log 5.9} = \frac{0.602}{1.229} = 0.49$$

or exactly the same result as in the previous case. This indicates that the result is fairly reliable. The agreement is less exact in other instances, but in most cases where three-stage survival could be used to check the estimates based on two-stage survival, the two estimates were fairly close to each other. In a few instances, four-stage survival could be used.

HALF-LIFE AND RATE OF EVOLUTION

Table 9.2 records the values obtained by this method. There is a considerable variation in the half-life estimates, but they show a tendency which may be summarized in the distribution in table 9.3. The determinations for the Neogene have a mode in the 0.40 – 0.49 class and form a skewed distribution, with the great majority of estimates falling between 0.3 and 0.6. All the Paleocene estimates are lower, and fall between 0.10 and 0.19. Probably a substage division of the Neogene sequences would yield some estimates falling in this class also, but the modes for the Paleocene and Neogene would nevertheless be quite distinct. The difference in rate of evolution is also evident from the separate items in table 9.2, and is set forth diagrammatically in fig. 9.2. Though part of the difference may be due to the less complete fossil record in the Paleocene, it probably also reflects actually higher rates of evolution in the Paleocene.

In the Neogene faunas as a whole, those of the Spanish Mio-Pliocene show the lowest rates of evolution (longest half-lives) and those of the Chinese Plio-Pleistocene the highest (shortest half-lives); the Siwaliks Plio-Pleistocene is intermediate. The difference may be due to sampling

Table 9.2.
Estimates of stratum percentages in faunas 1, 2 or 3 units apart, and of species half-life (with age as time unit).

	Stratum percentage at temporal distance[a]			Half-life in ages from temporal distance		
	1	2	3	1	2	3
Insectivora						
Spain	49.00	25.00	—	0.97	1.00	—
Siwaliks	—	—	—	—	—	—
China	—	—	—	—	—	—
Above combined	34.00	16.00	—	0.64	0.76	—
Paleocene	5.80	—	—	0.12	—	—
Chiroptera						
China	27.00	18.00		0.53	0.81	—
Primates						
Spain	—	—	—	—	—	—
Siwaliks	8.90	14.00	—	0.29	0.70	—
China	29.00	—	—	0.56	—	—
Above combined	12.00	9.50	—	0.33	0.59	—
Paleocene	8.70	—	—	0.14	—	—
Carnivora						
Spain	24.00	5.90	—	0.49	0.49	—
Siwaliks	11.00	1.60	—	0.31	0.34	—
China	17.00	2.50	—	0.39	0.38	—
Above combined	18.00	3.30	—	0.40	0.41	—
Paleocene	7.40	—	—	0.13	—	—
Glires						
Spain	28.00	4.20	—	0.54	0.44	—
Siwaliks	17.00	—	—	0.39	—	—
China	13.00	—	—	0.34	—	—
Above combined	18.00	1.16	—	0.40	0.31	—
Condylarthra						
Paleocene	14.00	—	—	0.18	—	—
Perissodactyla						
Spain	14.00	—	—	0.35	—	—
Siwaliks	42.00	31.00	—	0.80	1.03	—
China	32.00	—	—	0.61	—	—
Above combined	29.00	8.80	—	0.56	0.57	—

Table 9.2. (Continued)

	Stratum percentage at temporal distance[a]			Half-life in ages from temporal distance		
Artiodactyla						
Spain	25.00	61.00	—	0.50	0.50	—
Siwaliks	24.00	7.10	—	0.49	0.52	—
China	8.00	—	—	0.27	—	—
Above combined	19.00	4.40	—	0.42	0.44	—
Proboscidea						
Spain	56.00	37.00	36	1.19	1.39	2.03
Siwaliks	21.00	13.00	—	0.44	0.68	—
China	15.00	—	—	0.37	—	—
Above combined	31.00	18.00	8.9	0.59	0.81	0.86
Total fauna						
Spain	29.00	9.75	3.45	0.56	0.60	0.62
Siwaliks	23.00	9.30	—	0.47	0.58	—
China	14.00	1.03	—	0.35	0.30	—
Above combined	21.40	5.92	0.82	0.45	0.49	0.43
Paleocene	7.30	—	—	0.13	—	—

[a]The time unit is here one age in all cases except for the Paleocene, where it is ½ age.

Table 9.3.
Distribution of half-life estimates. Data from table 9.2.

	Frequencies	
Half-life	Neogene	Paleocene
0.0–0.09	—	—
0.1–0.19	—	5
0.2–0.29	2	—
0.3–0.39	12	—
0.4–0.49	14	—
0.5–0.59	12	—
0.6–0.69	5	—
0.7–0.79	2	—
0.8–0.89	4	—
0.9–0.99	1	—
1.0–1.09	2	—
1.1–1.19	1	—
1.2–1.29	—	—
1.3–1.39	1	—
—	—	—
2.0–2.09	1	—

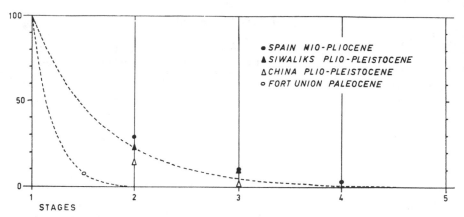

Fig. 9.2. Half-life curves fitted to the Neogene faunas of Spain, Siwaliks, and China (one average curve), and to the Paleocene Fort Union and Tiffany faunas.

chances and differences in taxonomic treatment, but it may also reflect accelerating evolutionary rates as the Quaternary influence becomes effective (the Chinese series goes farther into the Quaternary than the Siwaliks, whereas the Spanish one is entirely Tertiary).

The Insectivora, Chiroptera, and Primates have a relatively incomplete fossil record (the last-mentioned, moreover, are probably unnecessarily split up in the Siwaliks faunas), and the estimates are very variable, but little significance can be attached to this fact. The Carnivora have a good record, and the half-life estimates for Spain, Siwaliks, and China are relatively consistent, with an average half-life of about 0.4 ages. The record for the Glires (Rodentia and Lagomorpha) is less complete, and the values are somewhat more variable; but the differences are of doubtful significance. The average half-life is nearly the same as in the Carnivora.

The Perissodactyla of the Siwaliks appear to have been very long-lived in comparison with those of China and, especially, Spain; this probably reflects a real difference. The differences found in the Artiodactyla are probably also valid, but in this case the Siwaliks and Spanish forms give about the same values, whereas those of China have a shorter half-life. These facts may point to different evolutionary centres for the two groups; the Chinese high artiodactyl rate may, in addition, reflect the strong Plio-Pleistocene turnover of this order. It is interesting to note that the average half-life of the Artiodactyla is markedly shorter than that of the Perissodactyla (0.42 vs. 0.56), perhaps a reflection of the fact that the Plio-Pleistocene was the time when the even-toed ungulates underwent an "explosive" phase and rose to dominance.

The greatest differences are found in the Proboscidea; and they can hardly be due to biased data only. The half-life of the Spanish Mio-Pliocene proboscideans is radically longer than that of the Siwaliks and Chinese. Now, the Spanish sequence falls entirely within the time span of the mastodonts and deinotheres, probably by then to be regarded as old, well-adapted, and hence slowly evolving groups. There is the rather unusual spectacle of two species *(Turicius turicensis* and *Trilophodon angustidens)* persisting through four age-stages (Burdigalian to Pikermian, though not recorded in the Vindobonian). On the other hand, the Chinese series and the later Siwaliks record the rise and great deployment of the stegodonts and true elephants.

ESTIMATES OF MEAN LONGEVITY

There is as yet no very accurate age determination for the stages with which we have been concerned. However, the Tertiary may have had a length of some 70 million years, and if this is divided into 5 epochs with three ages each, it will be seen that the average length of a Tertiary age will be between 4 and 5 million years. Seeing that the data used here include the Villafranchian, which probably was shorter than 4 million years, and the Pliocene and Paleocene, which were probably below the average for epoch lengths, the value 4 million years may perhaps be used as the average equivalent of the Tertiary age.

The estimates of mean species longevity (in table 9.4) in the Neogene and Paleocene have been computed on this basis (half-life × 4,000,000 × 2.89, and rounded). From the previous discussion their highly tentative nature will be clear. The mean longevities for the Pleistocene have been taken from Kurtén (in the press).

The table suggests that the mean species longevity was much shorter in the Pleistocene and the Paleocene than in the Neogene, and that the order of magnitude does not differ much between the Paleocene and the Pleistocene; the values are roughly between 1/10 and 1/5 of the Neo-

Table 9.4.
Estimates of mean species longevity (in years) during the Paleocene, Neogene, and Pleistocene.

	Paleocene	Neogene	Pleistocene
Carnivora	1,500,000	4,600,000	740,000
Glires	—	4,600,000	460,000
Insectivora	1,400,000	7,500,000	320,000
Chiroptera	—	6,100,000	1,200,000
Condylarthra	2,100,000	—	—
Perissodactyla	—	6,500,000	—
Artiodactyla	—	4,900,000	—
Proboscidea	—	6,800,000	—
Primates	1,600,000	3,800,000	—
Total fauna	1,500,000	5,200,000	620,000

gene values. This would imply that rates of evolution were, on an average, between five and ten times faster in the Paleocene and Pleistocene than in the Neogene. Probably the acceleration reflects the fact that both the Paleocene and the Pleistocene were crucial times in the history of the mammals: the Paleocene, because the mammals were then surging to terrestrial dominance; the Pleistocene, because they were brought to bay by catastrophic environmental pressure.

REFERENCES

Colbert, Edwin Harris 1935. Siwalik mammals in the American Museum of Natural History. *Trans. Amer. Phil. Soc.*, new ser. 26:i–x, 1–401.

Crusafont Pairo, Miguel 1958. Endemism and Paneuropeism in Spanish fossil mammalian faunas, with special regard to the Miocene. *Soc. Sci. Fennica, Comm. Biol.* 18:1–31.

Kurtén, Bjorn, 1960. Chronology and faunal evolution of the earlier European glaciations. *Soc. Sci. Fennica, Comm. Biol.* 21(5):1–62.

Simpson, George Gaylord 1935. The Tiffany fauna, upper Paleocene. I. Multituberculata, Marsupialia, Insectivora, and Chiroptera. *Amer. Mus. Novitates,* 795:1–19.

Simpson, George Gaylord 1937. The Fort Union of the Crazy Mountain Field, Montana, and its mammalian faunas. *U.S. Natl. Mus. Bull.* 169:i–x, 1–287.

Teilhard de Chardin, Pierre, and Pierre Leroy. 1942. Chinese fossil mammals. A complete bibliography analysed, tabulated, annotated and indexed. *Inst. Geo-Biol. Pekin Publ.* 8:1–114.

TEN

The 'Half-Life' Concept in Evolution Illustrated from Various Mammalian Groups

INTRODUCTION

"PERCENTAGE DATING" has been known since Lyell defined his Cainozoic epochs on the basis of molluscan faunas and their relative content of modern forms. As a dating method it has been generally superseded by others and has long been only of historical interest. It is my task here to revitalize the concept and point to its ramifications in the study of evolution.

If we look at a living fauna (or flora), we might discern in it elements of greatly varying ages. Everybody knows the concept of "living fossils" which may have continued to exist for a hundred million years without perceptible change, but these are just extreme cases and the more commonplace creatures under our eyes may have very different histories. For instance, some of the mammals living at the present day in Europe were in existence, as the same species, some four million years ago, while others have arisen within the last half or quarter million years.

To appreciate the bearing of this fact on evolution and dating we must consider a given fauna, say the mammalian fauna of Europe, as it evolves through time. At a given moment the fauna contains a certain number of

Reprinted, with permission, from W. W. Bishop and J. A. Miller, eds. *Calibration of Hominoid Evolution* (1972), pp. 187–194. Edinburgh: Scottish Academic Press.

species: say 200. This is not a fixed number, of course, but in the very long run (before man meddles, at any rate) it tends to be approximately the same.

Three factors affect the composition of the fauna during the passage of time. One of them is immigration. What happens is that some animals, say in western Asia, find the conditions favorable and gradually spread farther and farther west. At the same time, some other animals already in the area may find the going a little tough. They do get on, but not quite so well as they used to. Their numbers thin out very slowly through the generations, until the moment comes when a male does not easily find a female to mate with, after which their extinction is quite a rapid affair. Extinction, then, is the second factor affecting the composition of the fauna.

These two factors counteract each other, and if I am right in saying that a fauna in any particular area tends to have approximately the same number of species when considered through time, it is evident that extinction and immigration will balance out over the millennia. Actually the two may be immediately correlated. An immigrating animal may have the same mode of life as one that was already there, but may be a little more efficient. If this is so, the old form will gradually die out. This is extinction as the result of competition.

The third factor in faunal change is evolution. A species evolves into another species which remains in the area, either continuing the way of life of its ancestor, or striking out into new avenues.

These, then, are the factors of change: immigration, extinction, and evolution. In a local fauna the first two factors might seem particularly important, but actually evolution is always there in the background. When a new species appears in a local fauna it will quite often be really new— i.e., specifically distinct from its ancestor. Thus the total change in a local fauna, which we may term the faunal turnover, should be very closely correlated with the rate of evolution of its components.

THE FAUNAL STRATUM AND ITS HALF-LIFE

The species constituting such a population at a given time will be termed the "stratum," or "temporal faunal stratum," of that time. If we assume that evolution is proceeding at a constant rate, the history of such a stratum will be of the type shown in figure 9.1. At an early date, only a few species belonging to this stratum are in existence and so they form only a small percentage of the fauna. As time passes and turnover continues, however, the percentage will increase. It will grow at a constant relative rate, but at an absolutely accelerating rate, and the curve depicting its rise will be a half-life curve like that in the figure. The half-life is the time during which the percentage is doubled—say from 20 to 40, from 50 to 100, etc. (See fig. 9.1)

When the percentage has risen to 100, we are at the given point in time after which the percentage of the stratum begins to diminish. As turnover continues, some of the species of the stratum vanish through extinction and some through evolution. If the rates of evolution and turnover remain constant, the percentage will fall off along a curve similar to that of its previous buildup, and we can again measure the half-life as the time of contraction from 100 to 50 percent and so on.

We can now see that the half-life concept is related to the theory of percentage dating. The faunal stratum that Lyell used was that of the present day and its percentages in fossil faunas show the gradual buildup described above. With absolute dates for one or more of the fossil faunas in hand, dates for the others can then be worked out through interpolation or extrapolation. Examples of this are given in Kurtén (1968).

The half-life is directly related to the mean longevity of the species in the fauna. The relationship between the two is

$$\text{Half-life/mean longevity} = \ln 2 = 0.6931$$

which, however, relates only to one of the two phases—the buildup or the contraction. The actual mean longevity will thus be twice as long as that obtained from this equation, or

$$\text{Mean longevity} = 2.89 \times \text{half-life}$$

Measuring species longevity is of great interest in the study of evolution. The difficulties, however, are great. In mammalian faunas, species are normally recorded as present in one stage, or in two or more. It is generally impossible to determine, even approximately, the longevity of species present in one stage only—and they are the majority. The species reported from two stages may actually be there throughout the two stages or only through parts of them. And, of course, part of the history of the species is often missing.

The half-life method, on the other hand, does not presuppose complete knowledge of the temporal range of the species. What we need is a succession of two or more large and evenly recorded faunas. If, for instance, two successive stages A and B are considered, the half-life may be determined, in terms of the age unit from the percentage, in the smaller fauna, of species present in the larger fauna. If three or more successive faunas are known, the results may be checked by counts of percentages in faunas at a distance of two or more stages. Finally, if we have absolute dates for the age-stages in question, absolute figures may be calculated for the half-life and mean longevity.

SOURCES OF BIAS

Various sources of bias affect the validity of the results attained in this way. There are three which seem especially important.

One is the incompleteness of the record. This will tend to reduce estimates of half-life. The importance of this factor is difficult to gauge but it is certainly radically lower for half-life estimates than for attempts to measure species longevity as such. It should be noted that the record will be better for comparatively large populations (common species) than for small, and there is reason to believe that such populations tend to be relatively conservative in evolution. This will counteract the influence of the bias.

Another important source of bias relates to problems of taxonomy. What constitutes a fossil species is still a matter of serious contention. Yet the species, in principle, should be more easily defined than any other taxonomic category; in neozoological taxonomy it is the only "objective" category. Unfortunately, as regards fossil mammals, the modern species concept has not yet been applied in all cases. Earlier students often used a typological species concept which led to over-splitting. This will reduce the apparent half-life. On the other hand, cautious students tended to refer incompletely known forms, tentatively, to a species already known; and this may tend to increase the apparent half-life.

Finally, it should be remembered that the absolute chronology is still not established with precision. The order of magnitude of the Tertiary and Quaternary age-stages is known but many details have to be filled out. To overcome this source of bias, it may be useful to work out average turnover rates for longer periods where such errors will tend to cancel out.

HALF-LIFE OF TERTIARY MAMMAL SPECIES

Four faunal sequences were studied by Kurtén (1959). They were (A) Spanish local faunas from the Burdigalian-Vindobonian-Vallesian-Pikermian; (B) Siwaliks local faunas from the Chinji-Nagri-Dhokpathan-Tatrot/Pinjor (combining the two last-mentioned into a single stage); (C) Chinese local faunas from the Pontian-Astian-Villafranchian-Mid Pleistocene; and (D) North American Paleocene faunas from the Fort Union and Tiffany. Example (C) carries into the Pleistocene proper and (B) into the Villafranchian, but the Villafranchian is a stage of Tertiary rather than Quaternary type.

The material was also broken down taxonomically into orders, in this case Insectivora, Chiroptera, Primates, Carnivora, 'Glires' (combining Rodentia and Lagomorpha), Condylarthra, Perissodactyla, Artiodactyla, and Proboscidea. The results of the study are summarized in table 10.1, in which the half-life is recorded with the average age-stage as a unit.

The Old World sequences used here start with the early Miocene, about 25 million years ago, and end with the middle Pleistocene about 1 million years ago. These 24 million years may be divided into about seven ages, so that each of the ages, on an average, lasted 3.4 million years. The average length of a Paleocene age is probably the same, seeing that the epoch as a whole probably lasted about 10 million years (Evernden et al. 1964).

HALF-LIFE OF PLEISTOCENE MAMMAL SPECIES

A study of faunal turnover in the Pleistocene mammals of Europe was carried out by Kurtén (1968) and it was shown that the rate of the turnover was accelerated during the Pleistocene, being considerably higher in the middle and late Pleistocene (post-Cromerian times) than in the early Pleistocene (Villafranchian). The half-life of the Villafranchian fauna was found to be about 1.5 million years, while that of the post-Cromerian faunas would be only about 400,000 years. In different orders, values varying from 180,000 to 1,600,000 were established for the relatively short time span from the Holsteinian Interglacial to the end of the Last Glaciation (but not including the mass mortality of the megafauna at the end of the Pleistocene).

Table 10.2 summarizes data on the half-life of various mammals in the Tertiary and Quaternary. Tertiary values are calculated from the data of Table 10.1.

Looking first at the average for the total fauna, we may note that the half-life in the Paleocene and late Pleistocene was only about one-third of that in the late Tertiary. This apparently reflects a very marked increase of the tempo of evolution both in the Paleocene and the later Pleistocene. The Paleocene phase of rapid evolution reflects the scramble for unoccupied niches in nature; that of the Pleistocene may be a reaction to the greatly increased environmental pressure of the Ice Age with its rigorous climatic zonation and rapid changes between interglacial and glacial conditions.

ORDINAL RATES OF EVOLUTION

The length of the half-life may also give information on the relative rates of evolution within different orders and in different epochs, and the data of table 10.2 are instructive in this respect. As regards the small mammals, Insectivora, Chiroptera and Rodentia, as well as the Primates, the incompleteness of the record is probably still an important factor of bias. Personally I think that the values given here (except for the Primates) mostly tend to be too long because only the better-known, more populous, and hence probably more conservative species have been included.

On the other hand, in the case of the Carnivora, the three ungulate orders, and the Proboscidea, the record is probably quite good and the values fairly realistic. In all of these groups, evolution was slower in the late Tertiary than in the Pleistocene (and the Paleocene for the groups then present). In no case is the difference more striking than in that of the Proboscidea, where the rate of evolution appears to have been intensified by a factor of ten or more in the late Pleistocene. This, I

Table 10.1.
Half-life, in terms of geological age-states, for Tertiary mammalian faunas
(see text for details)

		Half-life
Spain	Insectivora	0.98
	Carnivora	0.49
	Glires	0.49
	Perissodactyla	0.35
	Artiodactyla	0.50
	Proboscidea	1.30
	Total	0.55
Siwaliks	Primates	0.70
	Carnivora	0.33
	Glires	0.39
	Perissodactyla	0.91
	Artiodactyla	0.50
	Proboscidea	0.56
	Total	0.52
China	Chiroptera	0.67
	Primates	0.56
	Carnivora	0.39
	Glires	0.34
	Perissodactyla	0.61
	Artiodactyla	0.27
	Proboscidea	0.37
	Total	0.33
Above combined	Insectivora	0.70
	Chiroptera	0.67
	Primates	0.46
	Carnivora	0.40
	Glires	0.36
	Perissodactyla	0.56
	Artiodactyla	0.43
	Proboscidea	0.70
	Total	0.45
Paleocene	Insectivora	0.12
	Primates	0.14
	Carnivora	0.13
	Condylarthra	0.18
	Total	0.13

Table 10.2.
Half-life, in millions of years, for mammalian faunas at various times in the Cainozoic

	Paleocene	Miocene–Early Pleistocene	Late Pleistocene
Insectivora	0.4	2.4	0.5
Chiroptera	—	2.3	1.6
Primates	0.5	1.6	0.2
Carnivora	0.4	1.4	0.6
Rodentia	—	—	0.5
Lagomorpha	—	1.2	0.6
Condylarthra	0.6	—	—
Perissodactyla	—	1.9	0.3
Artiodactyla	—	1.5	0.5
Proboscidea	—	2.4	0.2
Total	0.44	1.5	0.54

think, reflects the fact that the Miocene-Pliocene was the climactic time of the mastodonts, which were by then an old, well-adapted, and slowly evolving group. In contrast the Pleistocene witnessed the rise and rapid deployment of the true elephants that replaced their ancient predecessors.

PERCENTAGE DATING

The half-life concept was used in percentage dating by Kurtén (1960a, b), demonstrating that not only the Recent faunal stratum, but also that of any other point in time, may be used for the purpose.

For instance, Locality 18 at Choukoutien has a mammalian fauna containing 10.2 per cent Astian species, 49.0 per cent Elster-glacial species and 10.2 per cent Recent species. The low incidence of Recent species indicates a Villafranchian date; in Europe the percentage of Recent species in early Villafranchian fauna is 6.7 and in late Villafranchian 10.9. The incidence of Astian species is also quite low and this would suggest a late date in the Villafranchian, which is consistent with the Recent percentage.

Finally, the relatively high incidence of Elster-glacial species suggests a date not preceding the Elster by more than about one half-life of the fauna. The age tentatively assigned to the Elster is about 400,000 years and the half-life probably intermediate between 1.5 and 0.4 million, say one million. This would give a date of about 1.5 million years for Locality 18. This is equivalent with the late Villafranchian, or roughly the age of the Tiglian Interglacial. The results are thus quite consistent.

The validity of such conclusions should, however, always be tested with the help of index fossils, radiometric dates and other dating methods as available. Percentage dating is just one method among many others, but one that has been somewhat neglected.

REFERENCES

Evernden, J. F., D. E. Savage, G. H. Curtis and G. T. James. 1964. Potassium-argon dates and the Cenozoic mammalian chronology of North America. *American J. Sci.* 262:145–198.

Kurtén, B. 1959. On the longevity of mammalian species in the Tertiary. *Comment. Biol. Soc. Sci. Fennica* 21(4):1–14.

Kurtén, B. 1960a. Faunal turnover dates for the Pleistocene and late Pliocene. *Comment. Biol. Sci. Fennica.* 22(5):1–14.

Kurtén, B. 1960b. An attempted parallelization of the Quarternary mammalian faunas of China and Europe. *Comment. Biol. Soc. Sci. Fennica* 23(8):1–12.

Kurtén, B. 1968. *Pleistocene mammals of Europe.* London: Weidenfeld and Nicolson; Chicago: Aldine.

V
PALEOETHOLOGY

ELEVEN
The Shadow of the Brow

Now, when any one with no covering on his head . . . strives to the utmost to distinguish in broad daylight . . . a distant object, he almost invariably contracts his brows to prevent the entrance of too much light.
  Charles Darwin, *The Expression of the Emotions in Man and Animals*

His hat, his plume, his sunburnt face, The shadow of his eyebrows' trace . . .
  J. L. Runeberg, *The Soldier Boy*
        (translated from the Swedish by C. W. Stork)

THE SUPRAORBITAL TORUS is found in all species of *Homo* (as well as in many other primates) except in *H. sapiens* of modern type. It is difficult to find any mechanical (architectonic) function for it. One possible and indeed probable function may be deduced from Darwin's (1872) remarks on the frown. He (and before him Gratiolet and Spencer) pointed to the eye-shading effect of the contracted brows. From such beginnings, then, the frown acquired a second and even more important role as an "expression of emotion" or, in ethological parlance, a threat display. Following Darwin's reasoning, von Haartman (1974) suggests that the torus originated as an adaptation to protect the eyes from direct sunlight in a steppe landscape.

The torus must have affected the visual image of the living face in a profound manner, as may be seen in various life reconstructions. We tend to read a menacing air into most of them, and we may suspect that some of the lay reluctance to accept an evolutionary origin of man stems not a little from such a subconscious influence. Did civilized human beings descend from such aggressive brutes? As Guthrie (1974:267) says,

Reprinted, with permission, from *Current Anthropology* (1979) 20:229–230.

"Even in humans, the exaggerated protruding bony brows of an old male gorilla connote awesome intimidation, which has made it easy for writers to portray it as a dangerous beast, even though it is a vegetarian and rather shy."

What we experience as an "intimidation stare" perhaps reaches its acme in the eagle, where, as Lorenz (1943) shows, the effect is due to a combination of two eliciting key stimuli: the shading bar over the eye, which casts a shadow like a deeply pulled brow, and the apparently tight-lipped "grimly" closed mouth. In the case of the eagle, these are "superstimuli," but, as von Haartman has shown (1974), it is this kind of picture that springs to mind from the poet's description of the soldier boy's dead father (and far more effectively in the Swedish original: *Hans hatt, hans plym, den bruna hyn/ Och skuggan från hans ögonbryn*).

The ethological significance of structural traits in fossil man has been little explored. Guthrie (1974) has provided us with an excellent discussion of status-enhancing intimidation characters in modern man. Among these, visibly protruding structures (especially in males), e.g., nose, chin, beard, penis (see also Eibl-Eibesfeldt 1973), play a prominent part. It is interesting to see that such features are often stressed in Palaeolithic art, and here not only with reference to males: the exaggeration of female characters in "Venus" figures may also suggest status-enhancing effects. As noted by both Guthrie and Eibl-Eibesfeldt, the social-signal function of genitalia and secondary sexual characteristics has tended to be overlooked.

It is easy to see that the supraorbital torus of say, Neandertal man, in combination with appropriate facial display, produces a highly effective intimidation stare, as is pointed out by Guthrie (1974:267). While agonistic to hostile strangers, the display would presumably have a reassuring effect on dependents (the soldier boy warming to his father's gallant image). However, the shaded eyes might also assist other kinds of mimical display. A well-known Neandertal reconstruction by McGregor (1926, profile: see discussion in Kurth 1958:222) seems to emanate respect-inspiring sagacity with a touch of the wily. Further, the shading of the eye might assist in disguising the exact direction of the stare (see Guthrie 1974:269 for a discussion of the ethological significance of an analogous case). No doubt additional suggestions may be put forward.

On the other hand, it is evident that the torus imparts a certain stereotypy to the facial expression, thus reducing the range of possible signals. This might give the condition in *H. sapiens* an adaptive advantage. It may be added that the torus-less condition is paedomorphic (the character is not present in small Neandertal children). As Lorenz (1943) has shown, paedomorphic traits are aggression-inhibiting and hence potentially adaptive in a species which had by then acquired a fairly efficient armory.

The supraorbital torus is present in both sexes, with only slight dimorphism. Its loss could perhaps imply a shift from a comparatively stereotyped adult-juvenile dichotomy to a more highly articulated male-female-juvenile trichotomy, with an increased differentiation of social roles.

Aggression by threat display is often a sham. The observation that the "grim and proud" eagle is in fact, to a great extent, a carrion-feeder (Lorenz 1943) is parenthetical in this connection, but Guthrie's observation on the shyness of gorillas is thought-provoking. Could it be that the disposition to aggressiveness is inverse to its apparent expression in display?

Whatever its significance, the reduction of the supraorbital torus in *H. sapiens* represents a completely new departure in the evolutionary history of our genus and suggests important ethological change. We are reminded of the "punctuated-equilibrium" theory of evolution so ably put forward by Eldredge and Gould (1972). According to the theory, significant evolutionary change is mainly associated with allopatric speciation, especially in small marginal populations. This favors the view of *H. sapiens* as a species distinct from torus-bearing hominids and points up the complexity of the *Gestalt* of hominid evolution discussed by D. Pilbeam at a 1978 Nobel symposium in Bofors. (Whether forms like Skhul man represent steps in the process of differentiation or result from interbreeding appears not to have been definitely settled.) Perhaps I should add that regarding *H. neanderthalensis* as a distinct species need not reflect in the least upon our view of this species as an exponent of highly evolved humanity. For all we know, Neandertal man may have surpassed *H. sapiens* in some of the traits we like to call specifically human. Yet he may have been a different species. As Pilbeam says, we need alternative scenarios.

REFERENCES

Darwin, C. 1872. *The Expression of the Emotions in Man and Animals.* London.

Eibl-Eibesfeldt, I. 1973. *Der vorprogrammierte Mensch: Das Ererbte als bestimmender Faktor im menschlichen Verhalten.* Vienna.

Eldredge, N., and S. J. Gould, 1972. "Punctuated equilibria: An alternative to phyletic gradualism." In T. J. M. Schopf, ed., *Models in Paleobiology,* pp. 82–115. San Francisco: Freeman.

Guthrie, R. D. 1974. Evolution of human threat display organs. *Evolutionary Biology* 4:257–302.

Kurth, G. 1958. "Betrachtungen zu Rekonstruktionsversuchen." In G. H. R. von Koenigswald, ed., *Neanderthal Centenary 1856–1956,* pp. 217–30.

Lorenz, K. 1943. Die angeborenen Formen moglicher Erfahrung. *Zeitschrift fur Tierpsychologie* 5:235–409.

McGregor, J. H. 1926. Restoring Neanderthal man. *Natural History* 26:288–93.

von Haartman, L. 1974. Skuggan från hans ögonbryn: En Runebergsstudie. *Societas Scientiarum Fennica Årsbok* 52:179–95.

INDEX

Pages containing illustrations are listed in *italics*.